Calcium Regulation
in
Sub-Mammalian Vertebrates

Calcium Regulation in Sub-Mammalian Vertebrates

Christopher G. Dacke

Physiology Department,
Marischal College,
University of Aberdeen,
Aberdeen AB9 1AS, UK

1979

ACADEMIC PRESS
LONDON NEW YORK SAN FRANCISCO
A Subsidiary of Harcourt Brace Jovanovich, Publishers

ACADEMIC PRESS INC. (LONDON) LTD.
24/28 Oval Road
London NW1

United States Edition published by
ACADEMIC PRESS INC.
111 Fifth Avenue
New York, New York 10003

Copyright © 1979 by
ACADEMIC PRESS INC. (LONDON) LTD.

All Rights Reserved
No part of this book may be reproduced in any form by photostat, microfilm, or any other means, without written permission from the publishers

Library of Congress Catalog Card Number: 78-67891
ISBN: 0-12-201050-7

Printed in Great Britain by
Latimer Trend & Company Ltd Plymouth

For Charlotte Elizabeth

Foreword

The publication of this monograph is extremely timely. Reviews have appeared from time to time dealing with calcium metabolism in specific vertebrate classes but there have been few, if any, devoted to all of the vertebrate classes. One exception was the very useful text by Professor Kenneth Simkiss entitled, "Calcium in Reproductive Physiology". In this text, published over a decade ago, the emphasis was on reproductive function in all vertebrate classes with the important exception of fishes.

It is appropriate that Dr. Dacke, Professor Simkiss' student, has chosen to update and enlarge the perspective and scope of this topic. He has included all the major sub-mammalian classes, especially fishes, expanded the perspective beyond reproductive function, and updated the summary of our knowledge in this area by including the very important developments which have occurred during the past decade, especially in the area of the endocrinology of Vitamin D.

Dr. Christopher Dacke is particularly well-equipped to review this area, and although the excellent monograph will speak for itself, it is pertinent to remind its readers that the author pursued his pre-doctorial training under Professor Simkiss at the University of Reading in avian calcium metabolism and was associated with Professor Chester Jones at the University of Sheffield where he gained valuable experience in electrolyte metabolism in fishes. In addition I am personally indebted to Dr. Dacke who spent two delightful post-doctorial years in my laboratory at the University of Missouri. He not only enhanced my understanding of the intricacies of avian calcium metabolism, but introduced me to working with fishes in collaboration with my academic colleague, Professor Robert Fleming.

There is no question in my mind that this text will prove invaluable to that increasing body of scientists interested in the endocrine and evolutionary aspects of the vertebrates. Important discoveries of relevance to man have been, and continue to be, made in the field of comparative endocrinology of the vertebrates. Without research in chickens and Japanese quail our understanding of vitamin D metabolism would be considerably poorer. Without fish studies, salmon calcitonin would not have become available for the treatment of Paget's disease of bone. Those of us, then, who are dedicated to studying the comparative aspects of vertebrate calcium metabolism, are to be indebted to Dr. Dacke for his labours in synthesizing and releasing this treatise, which, I am certain, will prove to be the standard work in its area for several years to come.

January, 1979

Dr. Alexander D. Kenny
Texas Tech University,
School of Medicine,
Lubbock, Texas

Preface

Calcium regulation in man and other animals has been studied for more than fifty years and the emphasis in most books and reviews dealing with this subject has been on the mammalian class. It has been my intention with the present book to review critically the relatively neglected subject of calcium regulation in vertebrate classes other than mammals, while, at the same time, presenting this from a personal and speculative viewpoint.

The book can be divided into two halves, the first of which deals mainly with the concept of calcium regulation and its evolution in early vertebrates. It considers the relationship between calcium and associated electrolytes in the environment and those in the body fluids, and then goes on to discuss the various hormones and target tissues involved in vertebrate calcium homeostasis. The second half offers an account of sub-mammalian calcium regulation on a class by class basis, ranging from the primitive, extant, jawless vertebrates to birds. Calcium regulation in vertebrates may be confounded by the stresses imposed during reproduction when calcium is transferred from mother to yolk, eggshell or foetal bone. It is also complicated by the need to regulate pH and other electrolytes such as magnesium and phosphate. A considerable part of the book is, therefore, devoted to discussion of these interesting topics and their interactions with calcium metabolism.

A final chapter of conclusions and speculations tries to relate the knowledge of calcium regulation in all vertebrates and to show how its evolution may have been affected by other homeostatic requirements. The emphasis in this chapter is more speculative than conclusive, which properly reflects the exponential growth of contemporary scientific investigation and the fact that each discovery usually leads to several new questions.

While I have tried to use an interdisciplinary approach, drawing upon experimental evidence from fields as diverse as palaeontology and biochemistry, the main viewpoint is that of a comparative physiologist. As it is impossible in a book of this length to discuss quantitative experimental evidence in great detail, a number of generalizations must, of necessity, be made. I have tried to balance this with as much hard data as possible in the form of figures and tables. Inevitably, with such an approach one has to be selective with the literature and often with more general or well-known aspects of the subjects, reviews have been cited in preference to individual papers. I apologize, therefore, to any authors who might feel that their contribution to the subject has been inadequately represented.

I am grateful to the many authors and publishers who have given permission to reproduce tables and figures and, in particular, to Dr. E. Lopez, Dr. D. R.

Robertson, Professor J. Yamada and Professor K. Simkiss who provided original photographs.

Special mention must be made of Ken Simkiss and Michael Radcliffe who read the whole of the first draft and to them, I am particularly indebted for criticism and encouragement. I must, however, take sole responsibility for the views expressed and the pattern of the book.

The figures were prepared by Alistair Simpson and the manuscript by my wife, Joan.

January, 1979   C. G. Dacke

Contents

Dedication	v
Foreword	vii
Preface	ix

1. Introduction	1
2. Calcium and Related Ions in the Environment and Body Fluids	6
The Environment	6
Body Fluids	10
3. Hard Tissues	12
The Choice of Calcium in Hard Tissues	12
Mineral Turnover in Hard Tissues	14
Mechanisms of Calcification	15
Precipitation	15
Inhibitors	15
Matrix	16
Ionic Activities	18
Cellular Mechanisms	20
4. Comparative Nature of Calcified Tissues	21
General	21
Bone	21
Apatite	21
Bone Fluid	22
Cells	23
Bone Formation	26
Acellular Bone	27
Fish Scales	29
Tetrapod Bone	31
Avian Medullary Bone	31
Calcified Cartilage	34
Calcium Deposits in the Inner Ear	35
Endo-lymphatic Sacs	36

Otoliths	37
Dentine	38
Other Calcified Tissues in Vertebrates	39
Soft Tissues as Calcium Stores	40

5. Calcium Regulating Hormones 41
   General 41
   Parathyroid Hormone 41
     Morphology 41
     Chemistry 43
     Pro-parathyroid Hormone 46
     Assay Methods 46
     Control of Secretion 48
     Circulating Parathyroid Hormone 49
   Calcitonin 50
     Ultimobranchial Morphology 52
     Chemistry 53
     Pro-calcitonin 57
     Assay Methods 57
     Control of Secretion 58
     Circulating Calcitonin 59
   Vitamin D_3 and its Metabolites 62
     Chemistry 62
     Control of 1,25-Dihydroxycholecalciferol Production and Secretion 65
     Circulating Levels of Vitamin D_3 Metabolites 66
   The Corpuscles of Stannius and Hypocalcin 67
     Morphology 67
     Nature of Active Substances from Corpuscles of Stannius 67
     Regulation of Stannius Corpuscular Hypocalcaemic Activity 68
   The Pituitary Gland 69
   Gonadal Hormones 70
   Other Hormones and Controlling Influences 72

6. Target Organs in Calcium Homeostasis 73
   General 73
   Bone 74
     Parathyroids and Bone 74
     Calcitonin and Bone 76
     Vitamin D_3 Metabolites and Bone 77
     Oestrogen and Bone 78
     The Pituitary and Bone 78
   Gut 79
     Parathyroid Hormone and Gut 79
     Calcitonin and Gut 80

Vitamin D_3 Metabolites and the Gut	80
Kidney	81
Parathyroid Hormone and Kidney	82
Calcitonin and Kidney	83
Vitamin D_3 Metabolites and Kidney	84
Prolactin and Kidney	84
Endolymphatic Lime Sacs	84
Skin and Scales	84
Skin	84
Scales	85
Gills	86
Oviduct	88
7. Calcium Regulation in Fish General Considerations	90
The Environment	91
Open and Closed Calcium Systems	92
8. Calcium Regulation in the Agnatha	94
Target Tissues and Blood Electrolyte Levels	94
Hormones	95
9. Calcium Regulation in the Chondrichthyes	96
Blood Electrolyte Levels	96
Target Tissues	97
Hormonal Influences	97
10. Calcium Regulation in the Osteichthyes	99
Blood Calcium Levels	99
Calcitonin and the Ultimobranchials	100
Calcitonin Injection	101
Ultimobranchial Extirpation	105
Plasma Calcitonin Levels	106
Hypocalcin and the Stannius Corpuscles	110
Stannius Corpuscle Extirpation	110
Replacement Therapy	113
Vitamin D_3 Metabolites	114
The Pituitary	116
Pituitary Extirpation	116
Replacement Therapy	119
Gonadal Hormones	121
Conclusion	122
11. Calcium Regulation in Amphibia	123
Plasma Calcium Levels	123

Parathyroid Hormone	125
Parathyroid Gland Extirpation in Anura	125
Parathyroid Injection in Anura	125
Lavage Studies in Anura	126
Parathyroids in Urodeles	126
Seasonal Cycles	126
Skin Response	128
Calcitonin and the Ultimobranchials	129
Ultimobranchial Extirpation	130
Calcitonin Injection	132
Vitamin D_3 Metabolites	134
General Effects	134
Effect on Gut	134
Lunar Influences	138
Pituitary Hormone	139
Other Hormones	140
The Endolymphatic Sacs and Acid-base Balance	141
Calcium Carbonate Reserves	141
Hormonal Influences	143
12. Calcium Regulation in Reptiles	**147**
Blood Calcium Levels	147
Parathyroid Hormone	149
Parathyroid Extirpation	149
Parathyroid Hormone Injection	149
Calcitonin	151
Ultimobranchial Anatomy in Reptiles	151
Calcitonin Injection	152
Vitamin D_3	152
Other Hormones	152
Reptilian Reproduction and Calcium Regulation	152
Hypercalcaemia	152
Calcium Reservoirs	155
13. Calcium Regulation in Birds	**156**
Calcium Regulation in Normal Birds	156
Parathyroid Hormone	156
Calcitonin and the Ultimobranchials	159
Vitamin D_3 Metabolites	163
Other Hormones	164
Calcium Metabolism in Egg-laying Birds	164
The Parathyroids in Egglay	166
Calcitonin in Egglay	169
Vitamin D_3 in Egglay	172
Acid-base Balance in the Egg-laying Hen	176

Calcium Appetite in the Egg-laying Bird 179
Conclusion 181

14. Conclusions and Speculations 183
   The Phosphate Problem 183
   The Magnesium Problem 187
   The Acid-base Problem 188
   The Calcium Problem 195

References 197
Subject Index 211

1. Introduction

In many animals, particularly vertebrates, calcium (Ca) is the most abundant cation in the body. This abundance generally reflects the presence of hard tissues containing an organic matrix impregnated with mineral salts, in which Ca is the major cation.

Calcium ions are important in many biological processes ranging from coenzymic functions to a role in the maintenance of structural integrity of biological membranes. They are essential in excitable tissues such as nerve terminals where they couple excitation to secretion, and muscle, where they are involved in the coupling of excitation to contraction. Calcium also has a general stabilizing effect in excitable membranes. A similar role for the cation can be found in the initiation of hormonal events including endocrine secretion, and later in the responses to hormones of their definitive receptors. It is often found in combination with carbonates or phosphates as a crystalline extracellular deposit. In these forms, Ca has a major structural function in the form of skeletons, scales, teeth, the protective coverings of cleidoic eggs and in the inner ear of vertebrates where calcified deposits may be involved in the sensations of hearing and balance.

There are two main lines of evolutionary pressures impinging upon Ca metabolism in animals. First we should consider the role of the divalent ion in dynamic physiological and biochemical processes. As these processes have become more complex, so this role has gained in importance. The increasingly complex interactions of many electrolytes in cellular activity has led to the need for a buffer compartment separating these cells from the vagaries of the environment. This extracellular or internal environment is relatively stable and provides a constant supply of electrolytes and other nutrients needed for the life of cells. As animals have colonized increasingly hostile environments such as fresh water and dry land, so the regulation of the internal environment has become more refined.

The second evolutionary influence which can be identified as affecting Ca metabolism concerns the development of mineralized skeletons. The most primitive of multicellular animals, the coelenterates and sponges, have a skeletal system based on an elevated turgor pressure within stiffened compartments (Istin, 1974). The turgor pressure in these compartments is based on two properties of fluids (1) that they are incompressible and (2) that they transmit pressure variations in all directions. In these hydrostatic skeletons a limited capacity for tissue contractility and movement is available. In their need to move in search of food or as a means of self-defence, many species

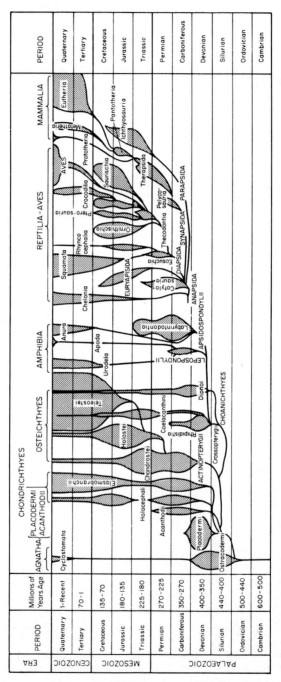

Fig. 1. A classification of vertebrates in relation to their phylogenic origins, and a time scale in terms of palaeontological periods. (From Torrey, 1971.)

INTRODUCTION 3

have developed more effective skeletons which are hardened by impregnation with mineral salts. These serve as rigid supports upon which contractile tissues may be anchored. Probably the first base for mineralization lay in the fibrous tissue surrounding the turgid compartments of hydrostatic skeletons. These hardened tissues would have developed into outer coverings for many animals, thus providing a double function, support for contractile tissues and protection against carnivorous predators.

In many species the mineralized tissues are located on the external surface. Increasing sophistication of this type of skeleton appears with the introduction of articulated joints which allow the mineralized tissues to act as levers; examples of such skeletons are found typically in the Arthropoda and Mollusca. They provide a solid protective covering but have the disadvantage of being relatively heavy and cumbersome and therefore limit the capacity of the animal to move. Vertebrates have overcome this problem by developing light and flexible skeletons based on the notochord. While not conferring the same protection from predators as the armour plating of exoskeletons, the possession of an endoskeleton allows an animal to move with relative speed and ease, this, in itself, affording a form of protection. Moreover, the endoskeletons of vertebrates have facilitated development of complex cranial structures and jaws, evolutionary trends which are a great advantage to both predatory and defensive modes of life. Thus the earliest known fossil vertebrates, the ostracoderms, had an external covering of bony armour, at least on the anterior surfaces. Later they began to develop internal bony skeletons while the exoskeleton was gradually reduced.

There are several theories concerning the evolution of bony tissues in vertebrates. The first recorded vertebrates were the ostracoderms of the Ordovician and Silurian geologic periods (see Fig. 1). These were jawless microphagous feeders related to modern cyclostomes, but unlike the latter extant order they had a bony armour referred to above. Romer (1964) and Urist (1976a) consider these early vertebrates to have evolved in a freshwater environment inhabited by large predatory scorpions, the eurypterids. According to Romer the bony armour of the vertebrates evolved as a defensive mechanism against these predators. Other workers, however, consider that the early vertebrates evolved in sea water (Halstead, 1974). Another older theory reviewed by Halstead (1974) suggests that bone first evolved as an excretory tissue. Seawater vertebrates are subjected to a constant hypercalcic challenge, but by depositing this ion into the skin it effectively becomes excreted. This excretory product might also have functioned as a Ca reservoir which could have helped the early vertebrates in their colonization of fresh water. A more recent theory proposed by Pautard (1961) suggests that bone evolved as a store for inorganic orthophosphate. Phosphorus as the phosphoryl moiety is part of important biologically active molecules such as ATP, DNA and RNA. The phosphate content of the sea is very low (see Table I in Chapter 2) and not in a form easily assimilated by animals. This ion can, however, be taken up by marine plant life (phytoplankton) which in turn are eaten by animals. There is a seasonal cycle of phosphate in the sea, since phytoplankton flourish in the summer and die during the winter. The phosphate which is returned to the sea is not readily available to animals, many of which also die at this time. Obviously, possession of a phosphate store would confer a major advantage upon a species.

Halstead (1974) considers that all the above theories may be, in part, correct and that bone could have served a number of purposes during its evolution. Thus it may have had a protective function (even if the predators were not freshwater scorpions), served as an excretory product for Ca and as a phosphate store, with the added advantage of acting as a rigid frame for the attachment of muscles.

As the role of bone as a mineral store developed (whether for Ca or phosphate) so the ability to regulate its turnover evolved. It is likely that hormones such as the vitamin D_3 system (undoubtedly hormonal) and calcitonin first made their appearance in response to this problem.

Rather surprisingly, in many extant aquatic representatives of lower vertebrates, the function of bone as a mineral store seems to have been at least partly lost. Bone is absent in the order Cyclostomata and very deficient in the class Chondrichthyes, these animals being left with only cartilaginous skeletons. It is generally believed that bone has degenerated in these types (Romer, 1964). In many bony fish, particularly the teleosts which form the great majority of living fish, bone has lost most of its cells and hence its ability to turn over. Perhaps modern fish have evolved other more sophisticated methods for regulating Ca and phosphate ions so that the function of bone as a mineral reservoir may be to some extent vestigial. In most vertebrates, including tetrapods, bone has, however, retained its cellular nature and would seem more likely to play a major role in Ca and phosphate turnover.

The evolutionary transition of vertebrates from sea water to fresh water and ultimately in their colonization of the land is undoubtedly reflected in the development of their Ca homeostasis. Thus the nature and complexity of the hormone systems and target organs associated with Ca homeostasis has changed and progressed in vertebrates and has probably assisted in their adaptive radiation.

It can be seen, then, that the Ca in skeletal mineral, body fluids and habitat are independent parts of a corporate identity and that considering the rapidity of Ca exchanges with the environment (at least in aquatic vertebrates), the corporate structure can be considered to be a continuum (Urist, 1976b). It is, therefore, with at least three exchangeable pools that Ca regulation in vertebrates is concerned. In fact there are several more pools if we take into account the intracellular environment, both cytosol and cell organelles, and more specialized compartments such as the bone fluid extracellular compartment and endolymphatic sacs (discussed in later chapters). The transfer of Ca and other electrolytes between these compartments is modulated by cellular membranes containing various types of energy dependent pump and diffusion mechanisms. In a species which spends its life entirely under stable environmental conditions such as sea water, the biological activity of the membranes may be pre-determined genetically. It is when conditions become variable, either due to changing environment or metabolic activity, that the activity of these membranes must be modulated by hormonal and, to a lesser extent, neural influences.

When considering the hormonal regulation of Ca metabolism, it is important to distinguish between mineral and skeletal homeostasis. Vaughan (1970) recognizes these as two distinct entities. She considers the hormones regulating Ca balance in the body fluids to comprise parathyroid hormone, calcitonin and perhaps the vitamin D_3 metabolites, while those regulating bone are represented by growth hormone, thyroid hormones and the adrenalcortical androgens. Other hormones which are important in the sub-mammalian vertebrates probably include prolactin, oestrogens and hypocalcin (the latter hormone only in the Actinopterygii). The possible role of these hormones as mineral versus skeletal regulators will be discussed in later chapters.

2. Calcium and Related Ions in the Environment and Body Fluids

The Environment

It is not disputed that life first evolved in the ocean. The earliest fossil prokaryotic cells date back more than 3000 million years and must have evolved under largely anoxic conditions. The constitution of these primitive oceans is not known with any certainty since different approaches to the problem of analysing them yield inconsistent results (Whitfield, 1977). However, it does seem clear that calcium (Ca) ions were present in oceans in relative abundance throughout their evolution. The first vertebrates probably appeared quite recently in geologic terms, about 600 million years ago during the late Precambrian or Cambrian era (Halstead, 1974). By this time the constitution of the oceans would have settled down to much the same as today. The composition of modern ocean is given in Table I and compared with that of the blood plasma in a primitive and advanced vertebrate.

TABLE I

Electrolyte composition of modern (Pacific) ocean compared with that in hagfish and human plasma

Component	Concentration (mmol litre^{-1})		
	Sea water	Marine cyclostome (*Eptatretus stoutii*) Plasma	Human plasma
Na	509	544	142
K	30	7·7	4
Mg	47·5	10·4	1·5
Ca	10	5·5	2·5
Cl	540	446	103
HCO$_3$	2·0	5·2	27
HPO$_4$	0·0	1·0	1·0
SO$_4^{2-}$	30	4·4	0·5
Si	0·02	—	—
pH	8·2	—	7·4
Total electrolytes	1169	1026·4	281·5

(Data from Urist, 1962 and Pitts, 1968.)

While Ca represents only a small fraction of the total ions in sea water, its concentration at 10 mmol litre^{-1} is still about four times that found in the extracellular fluid of most vertebrates. The Ca content of body fluids reflects the demarcation between invertebrates and vertebrates. In marine invertebrates the plasma Ca levels at around 10 mmol litre^{-1} are similar to that in sea water, while primitive vertebrates (cyclostomes) have plasma Ca levels of around 5 mmol litre^{-1} in the same environment (Urist, 1976b).

The other major cation in sea water is magnesium, the level of which is considerably higher than that in vertebrate extracellular fluid. Regulation of this important divalent cation has been relatively neglected. Its distribution around cellular membranes complements that of the Ca ion, at least in vertebrate cells, and it seems likely that magnesium metabolism is linked to that of Ca. When considering the evolution of Ca homeostasis we should bear in mind possible influences of magnesium and the need for its homeostasis. If primitive vertebrates were under constant challenge from high levels of environmental Ca, then they were certainly subjected to an even stronger challenge from high levels of environmental magnesium (see Table I).

Of the anions, phosphate and bicarbonate are the most important in relation to Ca metabolism. The concentration of the phosphate ion in sea water is extremely low; most of that available to animals would be in the form of tissues of plants or other animals. The vertebrates are the only class of animal to consistently use phosphates as the major skeletal anion, although it is found in the mineralized structures of some Protozoa, Coelenterata, Arthropoda and Brachiopoda (Pautard, 1961). Since phosphate in the ocean undergoes a seasonal cycle of availability (see Chapter 1), animals which possess a labile phosphate store are likely to be at an advantage compared to those that do not.

The bicarbonate ion is relatively abundant in the ocean and, not surprisingly, is incorporated into the mineralized tissues of many animals including those of vertebrates, in the form of $CaCO_3$ (see Istin, 1974). Apart from its availability as a source of the Ca anion, carbonate, bicarbonate is an important buffer. The pH of the ocean is mildly alkaline and free bicarbonate is available in such quantities that the ocean can be considered as a vast reservoir of buffer base. This has a bearing on acid-base balance in animals. Since acid is a by-product of metabolism most animals are faced with the prospect of controlling a more or less chronic metabolic acidosis. In sea water this is no great problem since bicarbonate anion is readily available and can easily be exchanged across tissues such as gills (Maetz, 1974); it is only when vertebrates began to colonize environments such as fresh water and, in particular, land, that acid-base balance may have presented serious problems. It is becoming increasingly apparent that the problem of acid-base balance has profoundly influenced the evolution of Ca metabolism in the vertebrates.

In nature there is a Ca cycle in which Ca may appear as one or other of its salts. The distribution of Ca and other elements in rocks and the sea is summarized in Table II.

In the rocks, Ca is found mainly in the form of natural or sedimentary phosphates, carbonates or sulphates. Over the geologic eras the natural phos-

TABLE II
Distribution of the principal elements in the lithosphere and hydrosphere

Lithosphere		Hydrosphere	
Element	% Weight	Element	% Weight
O	46·6	O	85·9
Si	27·7	H	10·8
Al	8·13	Cl	1·93
Fe	5·0	Na	1·07
Mg	2·1	Mg	0·13
Ca	3·6	S	0·09
Na	2·85	Ca	0·04
K	6·60	K	0·04
Ti	0·63	Others	0·01
P	0·08		

(After Istin, 1974, additional data from Urist, 1976a.)

phates have become dissolved by the action of acids so that Ca has entered the hydrosphere. The element has then been taken up by animals to form mineralized tissues. When these animals died, they sank to the ocean floor and formed sedimentary rocks which, in some cases, were raised and folded to form new hills and mountains, then eroded and redissolved by the actions of wind and rain. A diagram summarizing this Ca cycle is shown in Fig. 2.

Similar cycles exist for other elements. In the case of magnesium, the original deposits of this element in igneous rock would have been gradually eroded. However, since this element is not incorporated into the mineralized tissues of animals to any great extent, it does not precipitate from the ocean as rapidly as Ca. The main sedimentary rocks of magnesium are formed by evaporation or by combination with $CaCO_3$ to form complex dolomites. Thus while Ca is continuously recycled between the hydrosphere and lithosphere, magnesium tends to be recycled within the hydrosphere (Moore, 1972).

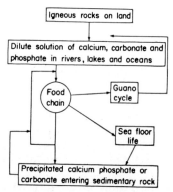

FIG. 2. Ca cycle in nature. (From Istin, 1974.)

This will account for the relatively high concentration of magnesium in sea water when compared with that of Ca.

The supply of phosphorus within the lithosphere, and hence the hydrosphere, is very meagre, and since this element is essential in living processes, it must be concentrated by tissues of plants and animals. About 95% of the phosphorus available in surface waters of oceans is used by organisms and eventually sinks as detritus to the deep ocean. About 1% of this is lost as sediment while the remainder is oxidized to inorganic phosphorus by biological action and returned to the surface by upwelling currents to repeat the cycle. The small percentage which is lost is replaced by new material fed in at the surface and the system is in a steady-state, although as mentioned in Chapter 1, seasonal fluctuations in phosphorus availability to higher animals will occur. This system in relation to other elemental cycles is reviewed by Whitfield (1977).

In fresh water, the electrolyte composition is much more variable than that of oceans, reflecting the different types of rock over which the rivers and streams are flowing. Some typical electrolyte values for hard fresh water are shown in Table III and compared with plasma electrolyte values in two

TABLE III

Electrolyte composition of fresh water (hard) from Lake Arrowhead, California, and freshwater vertebrate plasmas

Component	Concentration (mmol litre^{-1})		
	Fresh water (FW)	FW cyclostome (*Lampetra fluviatilis*)	FW teleost (*Coregonus elupeodes*)
Na	0·94	120	140
K	0·40	3·21	3·81
Ca	1·12	1·96	2·67
Mg	0·50	2·10	1·69
Cl	1·22	95·9	116·8
SO_4^{2-}	0·72	2·72	2·29
HCO_3	1·61	6·41	5·29
HPO_4	0	2·31	1·6
pH	7–9	—	—
Total electrolytes	6·51	235	276

(Data from Urist, 1962.)

species of freshwater vertebrates. The vertebrate species in this table are not necessarily found in the same fresh water as that for which analyses are given. While no electrolyte compositions of fresh waters can be said to be "typical" of all such waters, it is possible to make a number of generalizations. First, the total concentrations of solutes is much lower, usually less than 1% of that in sea water. Second, Ca generally replaces sodium as the most abundant cation. The Ca is usually a good deal lower in fresh than in sea water but it is normally present in sufficient quantities that animals living in such environments can be assured of adequate supplies.

It was noted by Rawson (1939) that the biologic activity of freshwater streams and lakes is related to their Ca content. Lakes containing less than 0·25 mmol litre^{-1} are graded as poor, 0·25–0·60 mmol litre^{-1} have medium productivity while 0·60–0·80 mmol litre^{-1} support a rich variety of both plant and animal life. The Ca in lakes may eventually form sediments which can then be raised by geologic upheavals and subjected to erosion and solution by rain in exactly the same way as the ocean deposits.

Magnesium is generally present at a lower concentration than Ca since there are relatively few sedimentary magnesium bearing rocks available for erosion.

Phosphorus is usually found in too low a concentration to be measured with accuracy. The pH of fresh water tends to be above 7·0 and the free bicarbonate content indicates an adequate reserve of buffer-base.

The first vertebrates to colonize fresh water might, at least as far as their Ca metabolism is concerned, have been confronted with an environment not markedly different from sea water in which they first supposedly evolved. The vertebrates would have faced a much more profound environmental change when they first crawled from water to land. In this situation, electrolytes such as Ca, magnesium and phosphate would have to be obtained entirely in the diet, or at least in drinking water, as an intermittent supply, rather than continuously diffusing across the gills and integument. The biggest problem of all, however, is likely to have been that of acid-base regulation induced by the loss of buffer-base reserve in the water. As we shall see, the development of acid-base regulation in tetrapods probably had a profound influence on their Ca metabolism.

Body Fluids

In man and other mammals the level of Ca in the blood plasma is regulated to within very close limits, usually between 2·25 and 2·75 mmol litre^{-1} (Simkiss, 1967). This generally holds true for the sub-mammalian vertebrates, although as we shall see in later chapters, the further down the vertebrate hierarchy one looks, the more variable are the plasma Ca levels. Most of the Ca in blood is found in the plasma and only small quantities are found in the erythrocytes according to Ohehy et al. (1966). Plasma Ca exists in two main fractions which are characterized as diffusible and non-diffusible. The diffusible fraction, which is thus named by virtue of its ability to pass through a dialysis membrane, can be further subdivided into a major ionized fraction and a smaller fraction which forms complexes with inorganic anions such as phosphate or citrate. Within the last few years it has become practicable to measure the ionic fraction of Ca in plasma directly by means of specific ion electrodes (Radde et al., 1971). Approximately 20–30% of the Ca in mammalian plasma is bound to protein and will not pass through a dialysis membrane (Vaughan, 1970). The values for sub-mammalian vertebrates are generally similar except in adult females of classes Osteichthyes, Amphibia, Reptilia and Aves during the reproductive season. In these classes the protein-bound fraction of Ca may be greatly increased due to the presence of yolk proteins which are synthesized by the liver (Simkiss, 1967); this aspect is dealt with in later chapters. The protein-bound Ca fraction is thought to be in simple equilibrium with the ionized fraction so that any change in the latter fraction will be reflected by the former. The level of plasma Ca can be affected by a number of non-hormonal factors of which the concentration of inorganic phosphate, protein, magnesium and pH appear to be the most important (Kenny and Dacke, 1975). Any changes brought about in plasma Ca levels by hormones are likely to affect the other electrolytes mentioned above so their association is clearly an intimate one.

3. Hard Tissues

The Choice of Calcium in Hard Tissues

As we have seen, calcium (Ca) is the most abundant of alkaline earth metals in the general environment, although in the sea, magnesium is more abundant. By far the majority of mineralized tissues in animals are composed of Ca salts which are also found in a few plant tissues. Other mineral forms are occasionally found in living tissues, for example the anionic silica skeletons of diatoms or the strontium sulphate skeletons of radiolarians (see Table IV).

In the majority of animals the major Ca salt in mineralized tissues is the carbonate, perhaps reflecting the relative abundance of bicarbonate in the environment. Only in the vertebrates and a few invertebrates is phosphate important, with formation of the complex Ca hydroxyapatite. Calcium carbonates are found even in vertebrates, both within the bone and elsewhere in specialized tissues such as otoliths (see Chapter 4). Mineralization of tissues, at least in multicellular organisms, always occurs within an organic framework consisting of a macromolecular structure. In animals this macromolecular framework consists of complex proteins and related substances which arise from the interlinkage of smaller polymeric units.

What are the factors which determine the choice of Ca as the major cation of mineralized tissues? To understand this we must look at the physicochemistry of this element and this has been reviewed by Williams (1970) and

TABLE IV
Examples of mineralized biological tissues

Species	Mineralized tissue	Chemical formula	Major organic component
Plants	Cell wall	$CaCO_3$	Cellulose, pectin, lignin
Diatoms	Exoskeleton	Si	Pectin
Radiolaria	Exoskeleton	$SrSO_4$	—
Coelenterates, Sponges and Molluscs	Exoskeleton	$CaCO_3$	Proteins, conchiolin
Arthropods	Exoskeleton	$CaCO_3$	Chitin, proteins
Vertebrates	Endoskeleton	$Ca_{10}(PO_4)_6OH_2$ and $CaCO_3$	Collagen, mucopolysaccharides

(From Istin, 1974.)

Istin (1974). The physiological role of Ca appears to be the result of a unique set of physico-chemical properties. First, it can form aqueous solutions in which the Ca ion is soluble; second, some Ca salts have a rather low solubility compared with those of ions such as sodium, potassium or even magnesium, and an equilibrium can exist between solid and dissolved forms; third, Ca tends to be excluded from cells while magnesium is retained by the action of ionic pumps located in the cell membrane. Thus, while the intracellular magnesium concentration ranges from about 2·5–22 mmol litre^{-1}, that of Ca ranges between 0·1 and 3 mmol litre^{-1} according to the cell type (Williams, 1970). If precipitates of Ca salts are to occur they tend, therefore, to be extra- rather than intracellular, although certain cell organelles such as mitochondria are an important exception to this rule (see below).

The stability of Ca salts is dependent upon a number of factors. The electron formula of the Ca atom is $3p^64s^2$ and consequently it can lose two 4s electrons to form a relatively stable Ca ion. The ionic potential, being the index of the strength of bonds which an ion can form, is determined by the ratio of the ionic charge (e) to its radius (r). Thus for ions such as magnesium and Ca where e/r is high, the ionic potential is also high, while for strontium and barium where e/r is lower, the ionic potential is lower (see Table V). The stability of ionic bonding depends both upon the ionic potential of the cation and of the anion and, consequently, the bond formation energy (E) of a purely ionic bond in any reaction will be represented by the cationic potential times the anionic potential.

TABLE V
Properties of alkaline earth ions

	Mg^{2+}	Ca^{2+}	Sr^{2+}	Ba^{2+}
Electron configuration	$2p^63s^2$	$3p^64s^2$	$4p^65s^2$	$5p^66s^2$
Atomic radius (Å)	1·36	1·74	1·91	1·98
Ionic radius (Å)	0·65	0·99	1·13	1·35

(From Istin, 1974.)

It can be seen that both magnesium and Ca ions are capable of forming relatively stable bonds. Another related property which helps determine ionic distribution in living tissues is that of isomorphic replacement between them when their radii differ by not more than 10%. Therefore, Ca ($r = 0.99$ Å) can be replaced by sodium ($r = 0.97$ Å) and provided the charge equilibrium is maintained, it cannot be replaced by magnesium ($r = 0.65$ Å). Connected with this property is the fact that Ca and sodium tend to be associated outside cells, although the association of magnesium and potassium inside cells is not obviously related to this property since the ionic radius (r) of potassium = 1·33 Å. However, the effective ionic radii can be altered according to the degree of hydration and the phenomenon of isomorphic replacement may not, therefore, be of great importance in determining ionic distribution around cell membranes. Another factor affecting the distribution of Ca in tissues lies in the property of ligand formation. When the water molecules surrounding a metal ion are replaced by other ions, metal complexes or coordination compounds are formed in which the group donating a pair of electrons to the metal ions acts as a Lewis base and is called a ligand. If the ligand (L) is a stronger Lewis base than water, water will be displaced from the hydrated metal ion $(M(H_2O))$ in the general reaction:

$$M(H_2O)n + L \rightarrow ML(H_2O)m + (n - m)H_2O$$

(n) and (m) refer to numbers of water molecules before and after ligand formation and follow standard nomenclature (Istin, 1974). If the ligand groups are attached to each other as well as to the metal, the metal can become part of a heterocyclic ring structure so that a metal chelate is formed. The sulphate residues, which are usually extracellular (for example in sulphonated polysaccharides), are linked to Ca rather than magnesium which has a much lower affinity for these groups and instead tends to bind with weak acid anions within cells (Williams, 1970; Istin, 1974; Bianchi, 1968).

Mineral Turnover in Hard Tissues

The precipitation of Ca salts into hard tissues has been the subject of much research and speculation. The mechanism by which new mineral is deposited into hard tissues is known as *apposition* and involves the *de novo* crystallization by Ca salts of newly formed collagen fibres, while that involving a net transfer of Ca ions into hard tissue matrix is known as *accretion*. Accretion may occur into an already existing hard tissue matrix but the physical chemistry of this process is probably similar to the mineralization stage in apposition. Removal of both matrix and mineral from hard tissues is known as *resorption*. There is some evidence that the two processes in resorption (at least in bone) might be separate (Vaughan, 1970), that by which net removal of Ca without matrix is accomplished being known as *calciolysis* (Raisz, 1976). The terms used above have been defined by Vaughan (1970) with respect to bone but it seems reasonable to use them when referring to other vertebrate mineralized tissues. While apposition must occur in all hard tissues, accretion and, in particular, resorption, are both associated with the remodelling of hard tissue and will only occur in mineralized tissues which retain the capacity for turnover. The mineral in such hard tissues is involved with electrolyte homeostasis in the extracellular fluid, while that in hard tissues which do not turn over is not. It is, therefore, with the exchangeable fraction of mineral in hard tissues that the various Ca regulating hormones are likely to be concerned. It is important to point out that only a small fraction of the solid mineral phase of bone is normally available for rapid physico-chemical exchange with the ions of the extracellular fluid; in human bone, for instance, between 1% and 5% of bone mineral is immediately exchangeable (Glimcher, 1976).

Mechanisms of Calcification

PRECIPITATION

The exchangeable fraction of mineral in hard tissues exists in dynamic equilibrium with that of extracellular fluids. Therefore, in order to understand the mechanism of calcification of the hard tissues, we need to have knowledge of the electrolyte composition of these extracellular fluids. How is it that the mineralized portions of these tissues are able to precipitate from a solution in which the concentrations of Ca and phosphate are about one half of the amount needed for spontaneous precipitation? If bone mineral is shaken with physiological saline, it dissolves and an equilibrium is obtained when the product of Ca and phosphate ions in solution reaches about 0·9 mmol litre^{-1}. This dissociation product is the point below which Ca hydroxyapatite will break down. If Ca and phosphate ions are added to physiological saline they will not begin to form a precipitate until their product reaches between 3·0 and 4·1 mmol litre^{-1}. The normal product of these ions in mammalian plasma is about 1·7 mmol litre^{-1} (Halstead, 1974), so that plasma is apparently undersaturated with regard to these ions and their precipitation does not occur. It is now recognized that bone mineral is formed not by precipitation, but rather by a process of crystallization at nucleation sites capable of lowering the energy barrier which would make precipitation possible at a localized site and which may, therefore, occur at much lower concentrations of Ca and phosphate than are required for direct precipitation. Recent data reviewed by Urist (1976) suggests that a physiological solution may induce formation of amorphous Ca HPO$_4$ with a Ca × P product as low as 0·63 mmol litre^{-1} and that it is only the presence, under normal conditions, of inhibitors, which are able to prevent crystallization within the general circulation and soft tissues.

The most critical stage of the crystallization process is its initiation, since, once formed, the crystals are able to grow from the metastable extracellular fluid surrounding bone. It is generally accepted that the fibrillar structure of collagen with its reactive sites, acts as a template for the initial seeding of apatite crystals. The type of collagen able to take on this role appears to be quite specific in that a native type 640 Å axial repeating structure is necessary. Furthermore, electron microscope studies suggest that the earliest detectable crystals are associated with the periodic banding along the collagen fibre (Vaughan, 1970). Collagen molecules from all vertebrate skeletal tissues appear to have remarkable structural homology according to Simmons (1976) who discusses them in more detail.

INHIBITORS

A further problem related to calcification of hard tissues concerns the mechanism by which mineralization of collagen molecules in extra-mineralized sites such as the connective tissues is prevented. A currently popular explanation suggests that calcification in these areas is counteracted by a strongly bound inhibitor, pyrophosphate (PPi) which is present in the plasma and urine (Fleisch, 1964). This hypothesis reviewed by Urist (1976), suggests that PPi, which is also present in bone, may play an important role in Ca homeostasis by either retarding the rate of bone mineral dissolution at resorption sites, or preventing Ca accretion by bone tissues. In bone cells there are two distinct pyrophosphatases, one with optimum activity under acid conditions, and one with optimum activity under alkaline conditions. Alkaline phosphatase can also act as a pyrophosphatase (Vaughan, 1970).

The magnesium ion may also act as an inhibitor of calcification. Leonard *et al.* (1972) have proposed that hydrolysis of calcium adenosine-5′-triphosphate (Ca^{2+}-ATP) initiates calcification and apatite formation. In non-calcified tissues this process does not occur and these authors have suggested that in such soft tissues where the concentration of magnesium exceeds the concentration of Ca, Mg^{2+}-ATP formation is favoured and apatite formation inhibited. Urist (1976) suggests that Ca HPO_4 nuclei are normally formed around the products of hydrolysis with Ca^{2+}-ATP, but with Mg^{2+}-ATP, Ca HPO_4 formation, and hence calcification, is inhibited. Magnesium in excess of Ca also tends to inhibit the hydrolysis of Ca^{2+}-ATP to yield inorganic phosphate and thus the formation of nuclei of Ca HPO_4. Furthermore, magnesium ions appear to increase the solubility of bone mineral. Perhaps this explains why primitive aquatic vertebrates of the order Cyclostomata do not form calcified skeletons since their plasma Mg/Ca ratios are very high (see Tables I and III). In other cartilaginous fish (members of the class Chondrichthyes and also of the order Chondrostei) the relationship between plasma magnesium and Ca levels appears to be much the same as in the bony Teleostei (Holmes and Donaldson, 1969; Urist, 1976b) so for these groups at least, the above theory is not corroborated. It is worth asking the question, however, has the refinement of a magnesium regulating system in higher vertebrates played a permissive role in the evolution of calcified tissues?

MATRIX

While collagen is undoubtedly the most important calcifying macromolecule, other organic macromolecules, which may act as nucleation sites, for calification exist. These include the keratins, the glycosaminoglycans and an as yet uncharacterized enamel protein (Halstead, 1974). It is suggested by Istin (1974) that these macromolecules are charged structures which create around themselves a micro-environment which is difficult to control. A local change in pH, for instance, will cause a considerable increase in concentration of one of the precipitating ions, thus creating conditions of hypersaturation. For a more detailed account of the organic constituents in bone, see Urist (1976a).

Glycosaminoglycans are found in bone as well as other sites and it is possible that the ability of bone and other tissues to calcify may depend partly on the presence of these substances (Vaughan, 1970). Indeed, evidence reviewed by Glimcher (1976) suggests bone glycosaminoglycans will bind Ca ions in preference to sodium, potassium or magnesium. Glycosaminoglycans may become attached to a protein core to form proteoglycans. Examples of glycosaminoglycans are the chondroitin sulphates which consist of glucoronic acid and acetylgalactosamine. The latter can be sulphated at the 4 or 6 position to give chondroitin-4-sulphate and chondroitin-6-sulphate respectively. It is of interest that chondroitin-6-sulphate is found predominantly in uncalcified cartilages, in sharks for instance, while chondroitin-4-sulphate is found in calcified cartilage in sharks and other species; in higher vertebrates it is the only chondroitin sulphate found in bone. Halstead (1974) feels that chondroitin sulphate is clearly associated with the process of calcification. The structures of chondroitin-4-sulphate and chondroitin-6-sulphate are compared in Fig. 3.

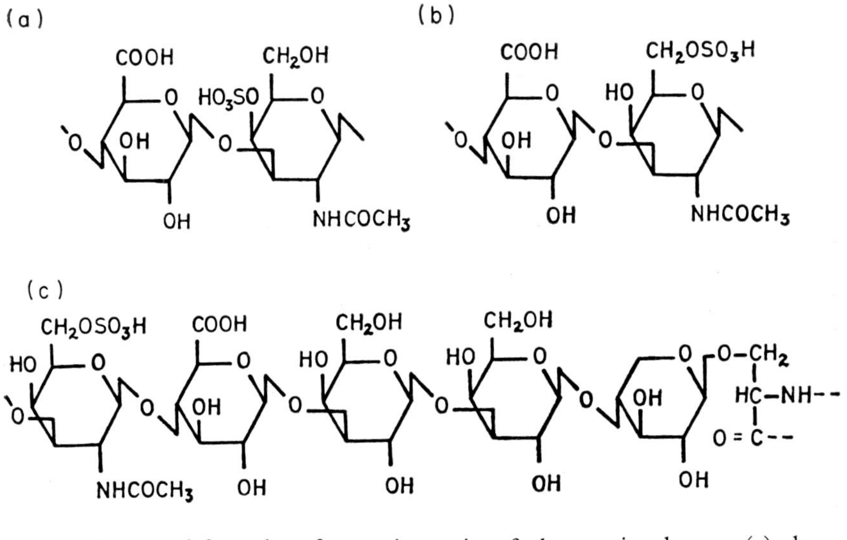

FIG. 3. Structural formulae of repeating units of glycosaminoglycans: (a) chondroitin-4-sulphate, (b) chondroitin-6-sulphate, (c) linkage region of chondroitin sulphate protein complex, reading from the left: chondroitin sulphate-galactose-galactose-xylose-serine. (From Halstead, 1974.)

Another substance currently receiving attention is an EDTA soluble non-collagenous protein rich in γ-carboxyglutamic acid (Gla). This protein has recently been identified in chicken as well as bovine bone matrix (mol. wt 20 000 and 6000 respectively). Glutamic acid, a newly identified amino acid, is isolated from alkaline rather than acid hydrolysates of EDTA extractable proteins. It was originally discovered in bovine prothrombin where it appears to play a permissive role in the Ca ion induced binding of prothrombin to phospholipid vesicles as well as binding insoluble salts such as $Ca(PO_4)_2$ (Urist, 1976a).

Ionic Activities

While the plasma of higher vertebrates may be considered to be metastable with respect to Ca salts, it is between two and four times more so in lower marine vertebrates and invertebrates than in bony fishes, according to Urist (1962). This is possible because the chemically active or thermodynamic ionic products are completely different from the apparent ion products. When the ionic strength of the extracellular fluids is high, the chemical reactivities of individual ions such as Ca and phosphate are reduced. Urist (1962) has calculated the ionic products for a number of physiologic plasmas and compared their ability to recalcify bone pieces from rachitic rats; these data are shown in Table VI. In species such as hagfish which have no calcified tissues the plasma Ca times inorganic phosphate ionic product may be 4·17 mmol litre^{-1} compared with around 1·23 mmol litre^{-1} for mammalian plasma, yet this product in hagfish plasma is not high enough to mineralize bone. The hagfish lives at a depth of 400 m, has a body temperature of 5 °C and the total ionic strength of the plasma is 0·70, close to that of sea water and about five times higher than mammalian plasma (0·16). These factors all combine to lower the effective ionic activities of Ca and inorganic phosphate. In the shark the Ca × HPO_4 ionic product is 4·43 mmol litre^{-1}, while body temperature is 15 °C; the total ionic strength is 0·29. This solution is apparently on the borderline of saturation with regard to tissue calcification since it can precipitate Ca salts into rachitic rat bone chips, but sharks do not, themselves, form stable calcified tissues on a large scale. It is difficult to compare accurately the saturation of blood plasmas from different vertebrate classes since knowledge of other relevant factors such as the activity coefficients, which are not presently available, is needed in order to calculate the thermodynamic ion products (Urist, 1962).

While extracellular fluids of all vertebrate classes may be saturated with respect to Ca salts, it does not follow that they are able to form calcified tissues; other factors which may be important include the presence of nucleation sites and certain enzymes. In sharks, for instance, plasma alkaline phosphatase levels are low, while in the matrix of calcifying cartilage they are high (Urist, 1976b). The possible role of this enzyme in the production of orthophosphate and removal of pyrophosphate has been mentioned previously. The presence of hormones which regulate Ca and inorganic phosphate transfer across various membranes is also an important factor.

TABLE VI

Chemistry of calcification

	Total ions (mmol litre^{-1})	Ionic strength (μ)	Total Ca (mmol litre^{-1})	UF[a] Ca (mmol litre^{-1})	Total inorganic P (mmol litre^{-1})	UF inorganic P (mmol litre^{-1})	UF Ca × UFP product (mmol litre^{-1})	Calcification of rachitic rat	Maximum (Ca × P)^3 of synthetic solution of same (μ)
Sea water	1168	0·77	11·0	10·8	0·01	0·01	0·11	0	—
Lobster	1180	—	11·3	—	0·02	—	4·52	—	—
Hagfish	1100	0·70	5·5	4·8	1·49	0·87	4·17	0	23·5
Shark	824	0·29	5·0	4·3	1·29	1·03	4·43	+	15·6
Teleost	340	—	3·0	1·8	1·93	1·29	2·23	+	6·5
Lamprey	235	—	2·0	1·2	4·19	—	—	—	—
Human	290	0·16	2·5	1·5	1·29	1·32	1·85	+	6·5
Fresh water, hard	6	—	1·0	1·0	0·01	0·01	0·01	0	—

[a] UF is ultra filtrable.

Cellular Mechanisms

While hard tissue calcification in vertebrates is essentially an extracellular phenomenon, it is obvious that intracellular processes are also involved. Evidence is accumulating to suggest that the seat of these intracellular processes lies in a variety of cellular organelles which include mitochondria, cytoplasmic RNA associated particles, matrix vesicles and matrix dense bodies (Urist, 1976a; Glimcher, 1976). These organelles may be involved in the translocation of Ca and associated ions across cell membranes and also possibly in the initial seeding of Ca salt crystals, a process which appears to take place in mitochondria. The activity of lysosomal enzymes may subsequently liberate these seeds to form the precursor or seeds for growth of extracellular apatite.

There are clearly several gaps still to be filled in our understanding of the mechanisms of hard tissue calcification in vertebrates or, for that matter, in all animals. It is not within the scope of this book to discuss these mechanisms in any great detail, but to describe them briefly in order to enhance our understanding of plasma Ca regulation. For recent detailed reviews of this topic the reader is referred to Urist (1976a, b) and Glimcher (1976).

4. Comparative Nature of Calcified Tissues

General

In this chapter a comparative description is given of the various types of calcified tissues to be found in vertebrates. This is brief, and for a more detailed account of vertebrate hard tissues, the reader is referred to Halstead (1974) and Simmons (1976) who give an account of the comparative physiology of bone.

While bone is obviously the most important of vertebrate calcified tissues with respect to Calcium (Ca) regulation, there are several other tissue species which may play a less important role in this process. These include calcified cartilage, dentine, aspidin, endolymphatic lime deposits and certain keratinized structures. Tissues such as fish scales, which may contain several elements of the aforementioned, will be considered, as well as a specialized calcified structure—the reptilian and avian eggshell.

Bone

APATITE

The evolution of a more or less ossified endoskeleton based on the notochord distinguishes the Sub-phylum Vertebrata from other animals. Yet in at least two major vertebrate classes, the Agnatha and Chondrichthyes, and also in a few of the Osteichthyes, ossified bone has either not evolved as a primary structure, or has become secondarily degenerate (see Chapter 1). Bone is still the major skeletal material in vertebrates, however, and like cartilage is a specialized form of connective tissue derived from the mesenchyme. It consists of transformed mesenchyme cells enclosed in a ground substance containing fibres of connective tissue. Laid down in this matrix are masses of crystalline salts with the general formula $Ca_{10}(PO_4)_6(OH)_2$. These correspond roughly to an hydroxyapatite, although the precise chemical structure of the bone mineral is not known. Approximately 99% of body Ca and 80% of phosphorus is found in the bone of higher vertebrates. There are also limited quantities of carbonates within the bone mineral, some of which are in the form of free carbonate, and some as part of a complex carbonate-apatite (Urist, 1976a). From studies of $NaH^{14}CO_3$ labelling in rats, it seems that bone "CO_2" is much more labile than either Ca or phosphate, but it is not clear to what extent this reflects free solution versus incorporation into bone mineral since the lability of the ^{14}C label when compared with ^{45}Ca and ^{32}P would not appear to be age dependent (Triffitt et al., 1968). There are small quantities of other ions such as magnesium, sodium, potassium and sulphate and trace minerals sequestered into the bone mineral (Vaughan, 1970). The mineral phase in mature bone constitutes 65% of the tissue wet wt (Glimcher, 1976). High concentrations of potassium are present in whole bone, but these appear to be in bone extracellular fluid rather than within the solid phase.

BONE FLUID

The anatomy of a possible bone extracellular fluid compartment is obscure. It is thought that bone fluid is separated from the systemic pool of extracellular fluid by a functional membrane which is able to exert an energy dependent control over the ion gradients (Neuman, 1969; Neuman and Ramp, 1971). The bone fluid has an unusual composition in that it contains relatively low Ca (0·5 mmol litre^{-1}) and high potassium (25 mmol litre^{-1}), at least in birds and mammals. Undoubtedly the bone "membrane" acts as an important target organ for Ca regulating hormones, although it is unclear whether or not this membrane is continuous with the various bone cell types discussed below. If whole freshly removed calvariae from neonatal rats are scraped on the outer surface and then dipped into physiological saline, 30% of the bone potassium

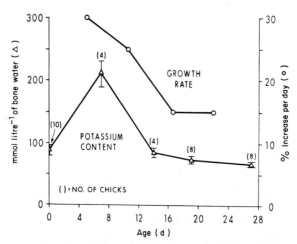

FIG. 4. Variation in bone potassium with respect to age of the chick ($\pm$ s.e.) and its correlation with rate of growth. Numbers in parentheses indicate the number of animals per point. (From Neuman, 1969.)

content is rapidly displaced. It is now thought that the bone membrane interface is continuous with osteoblast lining cells on the surface of bone (Talmage and Meyer, 1976). While the function, if any, of the high potassium content of bone extracellular fluid is not yet clear, it does seem to be related to the rate of bone turnover. Thus in vitamin D_3 deficient chicks, the bone sodium level is elevated from 24·2 to 28·1 mmol (100 mg dry wt^{-1}), while potassium is depressed from 74 to 50 mmol (100 mg dry wt^{-1}). These changes, which are reversed by vitamin D_3 therapy, are independent of plasma sodium and potassium levels (Neuman, 1969). The same author has demonstrated that high bone potassium levels correspond to a high growth rate in chicks (see Fig. 4) suggesting again a relationship between potassium and bone turnover.

CELLS

Enclosed within the bone matrix are cells, the osteocytes, which have an irregular stellate structure like the lacunae within which they become almost completely surrounded (see Fig. 5). They do, however, retain some tenuous cytoplasmic connection with each other via canalicules. Deeply seated osteocytes have the ultrastructure which one might associate with a resting cell. Hence, the cell organelles are poorly developed while the endoplasmic reticulum

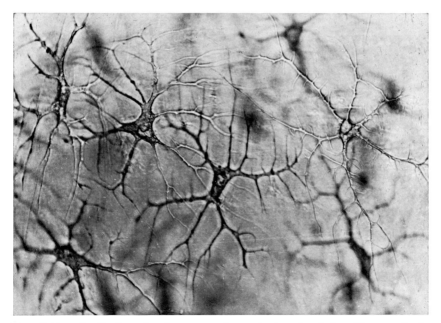

FIG. 5. Longitudinal section in a part of vertebral arch of *Anguilla anguilla* L. Osteocytes with ramified canaliculi. ($\times$ 280) (Courtesy of Dr. E. Lopez.)

is sparse and the mitochondria are densely granular (Vaughan, 1970; Halstead, 1974). There is, however, considerable evidence that osteocytes are involved in mineral homeostasis. It is thought that smaller osteocytes are involved in the continuous accretion of mineral into the bone while large osteocytes may be involved in bone resorption by osteocytic osteolysis (Raisz, 1976; Simkiss, 1967). More recent evidence indicates the osteocytes as the site of pumps which can rapidly move mineral to and from the bone surface (Talmage, 1972). The structure of the fish bone osteocyte (Fig. 5) is typical of that found in other vertebrate classes. The osteocytes are formed from mesenchyme cells which in an earlier phase are termed osteoblasts. Osteoblasts are involved in *de novo* bone accretion by secreting about themselves a thick matrix containing numerous collagen bundles into which a rapid deposition of Ca salts occurs. Once bone formation is complete, the osteoblasts become osteocytes. As this occurs, new osteoblasts are derived from the perivascular connective tissue (Osteoprogenator) cells along the developing edge of the bone (Simmons, 1976). Osteoblasts tend to have a columnar shape (see Fig. 6) with their nuclei furthest away from the developing bone surface. They do not have the ability to divide, but are replaced by constant recruitment from the mesenchyme (Fig. 6).

A third type of cell, the osteoclasts, are thought to be the major cell involved in bone resorption (Raisz, 1976). They are large and multinucleate, often with

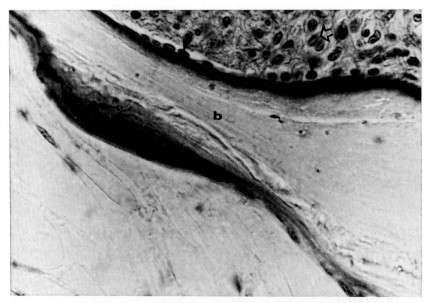

FIG. 6. Transversal section in a part of vertebral arch of *Salmo gairdnerii* R. Osteoblasts (▶), osteoprogenitor cells (⇒), bone (b). (× 252) (Courtesy of Dr. E. Lopez.)

several hundred nuclei. These cells originate by the fusion of precursor cells which are probably derived from blood-borne hematogenous elements—the monocytes (Simmons, 1976), and they are found along the surface of resorbing bone (see Fig. 7). Signs of osteoclastic activity are seen in the scalloped

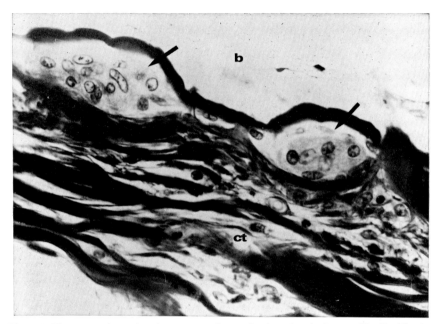

FIG. 7. Transversal section in a part of vertebral arch of *A. anguilla* L. Plurinucleated osteoclasts in Howship's lacunae (→). Bone (b), connective tissue (ct). (× 252) (Courtesy of Dr. E. Lopez.)

resorption pits (the lacunae of Howship). According to Halstead (1974) even the earliest fossil bones in primitive vertebrates show the characteristic signs of osteoclastic activity. Much of the substance of bone is laid down during its development in the form of layers (lamellae) which are found in adult bones, particularly in the surface layers. The morphological structure of bone is, however, often complicated by its continuous turnover. Thus in bone which has undergone remodelling, characteristic Haversian systems develop with small Haversian canals containing blood vessels as well as nerves branching through the bone. These canals are surrounded by concentric cylinders of bone containing cells which obtain their nutrients from the central canals. Formation of Haversian systems always occurs in remodelling bone which has undergone previous resorption; tubular channels are formed and bone is redeposited in characteristic concentric layers. Haversian systems are found in most classes of bony vertebrates, even fish in which parathyroid glands are lacking, but they are more common in large species. They are particularly associated with seasonal or circadian rhythms of Ca mobilization, or during periods of reproductive stress (Simmons, 1976).

The basic structure of bone outlined above is found in many extant vertebrate classes, and palaeontological evidence indicates that even the most primitive extinct vertebrates, the ostracoderms, had the typical structure discussed above (Romer, 1964). The most notable deviations from this general structure are found in modern teleost fish (see below).

Bone Formation

In vertebrates there are two quite different methods of bone formation or ossification to be seen in immature or embryonic forms (Romer, 1964). The simplest of these is formation of dermal or membrane bone, in which the ossified tissue originates directly from the mesenchyme without an intervening cartilaginous stage. This type of bone formation occurs in the skin or surface tissues. Groups of dermal mesenchyme cells take on osteoblastic characteristics and lay down between them thin membranes of fibrous interstitial matrix which then becomes mineralized. These plates are gradually enlarged and thickened by addition of bone layers on both inner and outer surfaces. In many bony fish, dermal bones are found over most of the body, including the mouth cavity, in the form of large plates anteriorly, and bony scales in posterior regions, where greater flexibility of the body is needed. The presence of dermal bone tends to be restricted in tetrapods. Bony scales disappear and in birds and mammals dermal bone is found only in the skull, jaws and pectoral girdle. The adult form of dermal bone usually consists of inner and outer layers of compact bone, with a more labile and obviously remodelled layer of spongy bone in between.

The second type of bone formation which is quite different and more complicated is that of endochondral bone. This involves replacement of an embryonic cartilaginous skeleton by an ossified structure and intracartilaginous ossification. Endoskeletal bones thus originate in the embryo as cartilaginous structures. In the centre of these "bones" the cartilage begins to degenerate and this process extends towards the distal ends of the bone where the cartilage cells begin to multiply and rearrange themselves in longitudinal rows. At first, this cartilage becomes mineralized, but it is then resorbed by penetrating blood vessels to form the bone marrow.

The ends of endochondral bones are not normally ossified until growth ceases, except in mammals and some reptiles (Simkiss, 1967). Long bones of vertebrates are thus hollow, consisting of thin tubes of rigid mineralized material. As every engineer should know, this type of structure gives a very advantageous weight to strength relationship. The advantage of this type of bone ossification compared with formation of dermal bone, is that mineral is not being merely added to a miniature preformed element, which by virtue of its rigid structure is incapable of fitting in with other elements and their complicated articular relations. Instead, the cartilaginous front is rather flexible and can easily be moulded to fit in with more elastic tissues. Thus Romer (1963) considers that while bone is the normal adult skeletal material in vertebrates (except in certain fishes and amphibians where osseous degeneration has occurred), cartilage is its indispensable embryonic auxillary.

It is possible for bone accretion and resorption to occur simultaneously at different sites in the same bone. A good example of this occurs in cultured chick foetal calvariae. These skull bones grow outwards by appositional growth on the outside and resorption of the inner surfaces. There is reason to believe that the two membrane surfaces, periosteum and endosteum, are performing different functions quantitatively, if not qualitatively (Newman *et al.*, 1973). Obviously remodelling of this type must be a complicating factor when considering the effects of regulatory hormones, particularly in respect of their actions on bone morphology.

The most marked differences in bone morphology when compared with the normal vertebrate pattern are found in fish.

In the primitive fossil dermal bone of ostracoderms, some placoderms and primitive osteichthyans, a spongy structure is seen with numerous spaces presumably filled with blood vessels. Star-shaped bone cell spaces are also present indicating the presence of osteocytes (Romer, 1963). No dermal scales or bones are found in degenerate cyclostome representatives, the lampreys and hagfishes; these have completely cartilaginous skeletons throughout their lives. In the Chondrichthyes the skin is practically naked except for the presence, in some instances, of fin spines of a dentine-like material and degenerate placoid scales or "dermal denticles" which are toothlike structures with an inner pulpy cavity and outer dentine and enamel covering. It is now considered that these scales represent the last vestige of ancestral dermal bone (Romer, 1964).

ACELLULAR BONE

The most widely spread and successful class of fish, the Osteichthyes, and in particular the order Teleostei, show a wide variety of bone types in that they may or may not contain osteocytes and are, therefore, considered respectively to be "cellular" or "acellular". Simmons (1971) has reviewed in detail the relationships between structure and function in teleost bone. Moss (1961) classifies teleost bone on the basis of whether or not it contains osteocytes. Cellular bone is most prevalent in the lower orders of the Clupeiformes and Scopeliformes. A few species, for instance the bonefish (*Albula vulpes*), have both cellular and acellular bone. Although fish acellular bone apparently contains no cells, its initial formation must depend upon the activity of cells of the osteoblastic or osteocytic type. The morphological picture of teleost acellular bone suggests a variety of ways in which it may be formed. Basically, these would involve either cellular withdrawal after mineralization, or a process of cellular self-burial which would need to be so complete as to totally occlude the lacunae and even the fine canaliculi in which the living cells had previously existed (Simmons, 1971). In teleost fish it is possible to trace the evolution from typical bone which contained osteocytes to that in which the cells are confined to vertical ascending passages termed "canals of Williamson" with only the branching cell processes extending into the bone tissues, and eventually, to a type in which the cell processes do not penetrate. Halstead (1974) considers there is no doubt that fish acellular bone has evolved from a basic cellular type.

A fundamental question, then, concerning teleost acellular bone, the major bone type of this order, is whether or not it is alive. Is this tissue labile like cellular bone, or is it metabolically inert and thus unable to contribute to the Ca homeostasis of the fish? It is not clear why most modern fish should have abandoned the ancestral cellular bone in favour of an acellular type, particularly if it is inert. Most available evidence suggests that acellular bone undergoes little remodelling or mineral exchange following fracture or dosage with radiocalcium, when compared with cellular bone (Simmons, 1971; Simkiss, 1974). The data of Moss (1962) are typical in this respect; he studied fracture repair in cellular boned goldfish and acellular boned cichlids (see Table VII).

TABLE VII
Repair of fish bone fractures under normal and hypocalcic conditions

	Treatment	*Carassus auratus* (cellular bone)	*Tilapia macrocephala* (acellular bone)
Normal water	Normal diet	+	+
	Acalcic diet	+	+
	Starvation	—	—
Acalcic water	Normal diet	+	—
	Acalcic diet	+	—
	Starvation	—	—

(From Moss, 1962.)

It is clear from these studies that cellular boned fish can repair fractures in all but the most hypocalcic of environmental situations, indicating a role for resorption of Ca reservoirs. Acellular boned fish can only repair fractures when a reasonable supply of environmental or dietary Ca is available. More recently Simmons et al. (1970), using the acellular boned toadfish (*Opsanus tau*), injected ^{45}Ca together with tetracycline and investigated the distribution of the isotope by microradiography and autoradiography. The results indicated that the radiocalcium label was retained mainly at sites of bone growth rather than being diffusely distributed within deeper bone as it would be in cellular bone. Furthermore, the ^{45}Ca at bone surfaces appeared to be unavailable for further exchange with the body fluids.

From these studies it would seem that Nelson (1967) had an appropriate concept of fish acellular bone as "a geobiochemical sink for mineral elements with little or no capacity for biologic turnover". However, Fleming (1967) considers that mineral can be withdrawn from teleost acellular bone, albeit at a very slow rate. He suggests that while this might have no role in Ca homeostasis, it could be related to the need for a continuous supply of phosphorus for metabolism, a possibility to be considered in some detail in Chapter 14. Recently Brehe and Fleming (1976) have studied seasonal changes in calcified tissue turnover in the acellular boned killifish (*Fundulus kansae*) and concluded that the Ca pool in the bone was exchangeable, with an influx half time of around 100 days in the summer months, falling to between 300 and 500 days in winter. The efflux half times ranged from 160 to more than 500 days, but these were rather variable with no obvious seasonal pattern. The corresponding half times for the labile Ca pools of the skin and scales were about a third of those for acellular bone. These data are also very interesting when considered in the general context of phosphorus metabolism.

In mammals, a form of acellular bone is the cementum surrounding teeth (Halstead, 1974); this is an analagous development to fish acellular bone. Presumably it is advantageous that bone in the vicinity of a tooth should not be readily available for resorption.

FISH SCALES

Fish scales are a modified form of the dermal endoskeleton of more primitive vertebrates. Most fish scales contain both dentine and bone in varying proportions and, therefore, represent a significant fraction of the bodily Ca reserves of these animals. The structure of fish scales has been reviewed by Van Oosten (1957) and more recently by Halstead (1974). Scales may be of three basic types. These are placoid scales, found in the Chondrichthyes, cosmoid scales, found in Crossopterygii, and ganoid scales which are found in the Actinop-

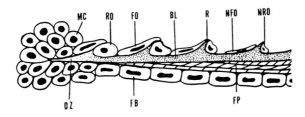

FIG. 8. Schematic diagram showing the structure of anterior margin of a teleostean fish scale and the scale-forming cells. Bony layer (BL), fibroblast (FB), flattened osteoblast (FO), fibrillary plate (FP), marginal cell (MC), necrotic flattened osteoblast (NFO), necrotic round osteoblast (NRO), osteoid zone (OZ), ridge (R), round osteoblast (RO). (From Kobayashi et al., 1972.)

terygii. Placoid scales consist mainly of dentine with an enameloid cap; in primitive sharks there may also be a basal bony layer. Cosmoid scales consist mainly of dentine and may have a hardened enameloid surface, while ganoid scales of actinopterygian fish generally have a basal layer of bony material termed "isopedin", covered by a layer of dentine. In advanced scales of teleosts, usually only the bony isopedin layer survives (Halstead, 1974). Scales of teleosts are further subdivided into cycloid and ctenoid types characterized by the form of annual growth rings. These scales consist of a bony (isopedin) layer deposited over a fibrous lamellar plate which does not calcify. There appear to be layers of osteoblastic cells associated with the lamellar layer, but not embedded within the bony layer to form osteocytes. Also, there are layers of fibroblast cells associated with the fibrous plate (see Figs 8 and 9). It is

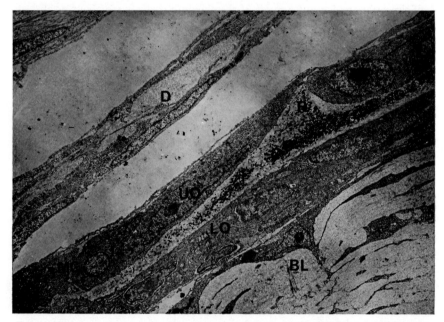

FIG. 9. Electron micrograph of saggital section through a goldfish scale. Ridge (R), ridge cell (RC), osseous layer (O), dermal scale pocket wall (D), upper osteoclasts (UO), lower osteoblasts (LO), basement lamellae of dermis (BL), marginal cells (MC), scale centre (→). (× 3360) (Courtesy of Dr. J. Yamada.)

not clear whether osteoclasts are present, but there is no doubt that these scales have the ability to turn over (Simmons, 1971; Simkiss, 1974). Thus the scales of migrating pre-spawning salmon and trout show evidence of scale resorption at a time when the fish cease to feed, with remineralization occurring when the fish return to the sea following spawning (Chrichton, 1935). Furthermore, evidence from Garrod and Newell (1958) indicates a fall in the

Ca content of around 20% of the scales during development of the ovary, with calcification recommencing after egg-brooding, and recently Mugiya and Watabe (1977) were able to induce scale resorption by injecting oestrogens.

The scales are well vascularized and, on a weight for weight basis, seem to have a greater capacity to take up Ca than the endochondral bone in fish. Thus Simkiss and Yarker (cited by Simkiss, 1974) found that the skin and scales contained about 40% of the total body Ca in the trout. When ^{45}Ca was injected, about 50% of the dose was retained in the skin, even after one week. Results from Mashiko et al. (1964) indicate that radiocalcium deposited in the scales of the crucian carp will be released, albeit slowly, if the fish are placed in water containing more than 60 p.p.m. CaCl, or alternatively, injected with CaCl.

These data, then, collectively indicate that in teleost fish, highly modified dermal bone in the form of scales, may have a greater capacity for turnover than endochondral bone, particularly of the acellular type. This suggests that scales in modern bony fish play a major role in Ca regulation.

TETRAPOD BONE

In tetrapods, dermal bone, in particular scales, are much reduced or nonexistent, while the endochondral bone becomes more important both as a supportive tissue and as a Ca reservoir. Primitive amphibians such as labyrinthodonts were well ossified, but modern anurans, urodeles and apodans show evidence of skeletal degeneracy while the endolymphatic sacs become prominent as mineral reservoirs (Romer, 1960). In some fossil amphibia, scales were present, but they are almost completely lacking in living amphibia except as functionless vestiges buried in the skin. Remnants of scales are also found in older extinct reptiles and in a vestigial form in the lizards and crocodiles. The skin still retains a potential for forming dermal bone. Examples of this are found in reptiles for instance, the crocodile armour of subquadrate bony plates covered by horny skin, and the massive bony shell of turtles (chelonians). In mammals examples of bony armour are found in edentate stocks of South America, the extinct ground sloths, and the armadillos (Romer, 1963).

AVIAN MEDULLARY BONE

In female birds, a very specialized form of endochondral bone is found in association with egg-lay. This is related to the fact that 25–40% of Ca in an eggshell is derived from the bones (Comar and Driggers, 1949). This bone is formed within the medullary canals and is termed "medullary bone". It was first described by Foote (1916) and later rediscovered by Kyes and Potter (1934), who noticed its coincidence with maturation of the ovarian follicle in female birds. Structurally, medullary bone consists of a mass of small intermeshed spicules which grow from the endosteal surface out into the medullary cavity giving a honeycomb appearance (see Fig. 10). The bone is well vascularized and has a large surface area for cellular activity. Indeed, during its formation and resorption there appear to be intense phases of osteoblastic and osteoclastic activity (see Fig. 11). These cyclical phases occur as successive eggs are laid. From histological evidence it has been suggested that osteoblasts

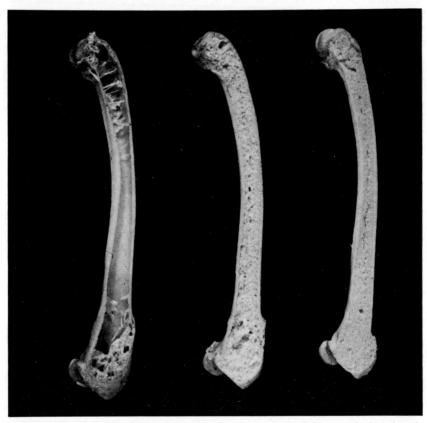

FIG. 10. The femurs of an adult male Japanese quail (left), egg-laying female (centre) and oestrogen treated male (right), split open and extracted with ethylene diamine to remove organic material. The endosteal surfaces of the laying hen and oestrogenized cockerel are covered with spicules of medullary bone which almost fill the medullary cavity.

in medullary bone may convert to osteoclasts and then back again. Thus there are few mitoses in these cells, but intermediate cell types are frequently found (Bloom et al., 1941). In avian medullary bone, then, osteoblasts and osteoclasts might be merely different functional states of the same cell type, although such a hypothesis should be treated with caution at present.

COMPARATIVE NATURE OF CALCIFIED TISSUES 33

The stimulus for medullary bone formation clearly originates from the ovary. Experiments with immature birds or male castrates indicate that both oestrogens and androgens are necessary for medullary bone induction. It is possible to initiate its formation in intact adult males by injections of oestrogens alone, although in immature birds, both sex hormones are needed (Bloom et al., 1940). It is not possible to promote medullary bone formation with oestrogen and androgen injection as quickly as can be seen under natural conditions during the pre-ovulatory period, at least in pigeons (Bloom et al., 1942), thus indicating a role for an additional factor or factors. In female birds, medullary bone is widespread throughout the skeleton, but in variable quantities. It is most obvious in limb bones where it may constitute up to 40% of the total weight (Taylor and Moore, 1954).

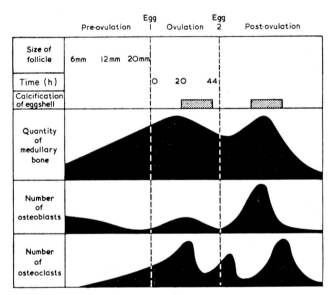

FIG. 11. Changes in the number of osteoblasts and osteoclasts in the medullary bone of laying pigeons in relation to the calcification of the eggshell. (From Simkiss, 1967.)

The organic component of medullary bone is different in many respects from that of cortical bone. For instance, medullary bone contains less than half the collagen on a dry weight for weight basis compared with cortical bone. Furthermore, the histological picture is more consistent with that of young, actively growing cortical bone and suggests that it can be rapidly formed and destroyed (Simkiss, 1967). He suggests that these differences are consistent with a role of medullary bone as a labile Ca store which can be utilized during periods of Ca stress (i.e. egg-lay) in female birds, rather than acting in any structural capacity.

Medullary bone seems to represent the opposite extreme from fish acellular bone in terms of bone evolution. It is surprising that reptiles such as crocodiles and turtles which also lay eggs with calcified shells, do not form medullary bone, which must, therefore, be an advanced evolutionary modification. The bone is not generally found in mammals.

Calcified Cartilage

Cartilage is the alternative major skeletal tissue found in vertebrates. It is particularly important in Cyclostomata and Chondrichthyes and also in the Chondrostei. In these three groups of fish, the ossified internal skeleton is almost completely degenerate, and is replaced by cartilage.

Typical hyaline cartilage is a flexible, elastic tissue. It has a ground substance or matrix of sulphated polysaccharide (proteoglycan) which forms a rather firm gel. Connective tissue fibres are found throughout this substance as well as cartilage cells, the chondrocytes, which are found in spaces (lacunae). Unlike the bone lacunae with their osteocytes, cartilage lacunae do not seem to be interconnecting. Blood vessels are generally absent so that nutrients must reach the cells by a process of diffusion through the matrix (Romer, 1963). There are no chondroclasts in chondrichthyan cartilage, and consequently, once formed, this tissue remains for the duration of the life of the individual (Urist, 1964).

Hyaline cartilage may become calcified. This occurs in long bones of higher vertebrates as the long bones ossify. The cartilaginous portion of the bone becomes calcified prior to its ossification. Cartilaginous skeletons of elasmobranch fish may also become calcified. A deposition of Ca salts into the matrix produces a relatively hard, brittle substance, something like bone. Calcification does not prevent nutrients from reaching the enclosed cells, rather there are zones of calcification with bands of uncalcified cartilage in between, allowing nutrients to penetrate and maintain the viability of the tissue. According to Halstead (1974), calcified cartilage was present in early fossil vertebrates, the Ostracodermi such as *Eryptychius,* indicating the primitive nature of this tissue. The cartilaginous skeletons of extant cyclostomes do not become calcified.

Experiments with radiocalcium indicate that virtually all the Ca in calcified cartilage is exchangeable and a simple equilibrium therefore exists between the extracellular fluid and the cartilage apatite (Urist, 1961). In a later review, Urist (1976b) makes a comparison of calcified cartilage with an ion exchange column for the Ca ion. This is in contrast to the situation in bone where only about 2% of the skeletal mineral is available for exchange. There is no evidence in the literature to suggest any hormonal role in the turnover of mineral in calcified cartilage of elasmobranchs. Shark calcified cartilage matrix contains high levels of alkaline phosphatase in contrast to the blood levels of the enzyme which are low. The role of this enzyme in the hard tissue physiology of sharks is just as obscure as in other vertebrates (Urist, 1976b).

Calcium Deposits in the Inner Ear

As we have seen, the major form of calcified tissue in vertebrates is hydroxyapatite, a complex salt based on Ca and phosphate. These salts are found principally in skeletal structures such as bone, scales and teeth. In many vertebrates, particularly higher tetrapods, more than 96% of the bodily Ca reserves are found in bone in the form of apatite. In the invertebrates, much of the Ca reserves are in the form of $CaCO_3$ which play a role not only as a skeletal support, but also in acid-base balance (Dugal, 1939). In vertebrates, small quantities of $CaCO_3$ may be found in bone. The quantities recovered from bone ash of representative species were: man 5·3%, turtle 13·1%, frog 1·5% and fish 0%. These bone deposits are usually in the form of carbonate-apatite or as aragonite crystals which are labile during periods of metabolic acidosis in man, and are, moreover, the first component of bone to be mobilized in this situation (Blitz and Pellegrino, 1969). In vertebrates, deposits of $CaCO_3$ are also found in the inner ear. In some species, particularly anuran amphibians, very substantial stores of these salts are found in association with the labyrinthine structures of the inner ear.

The membranous labyrinth consists of a series of sacs and canals contained

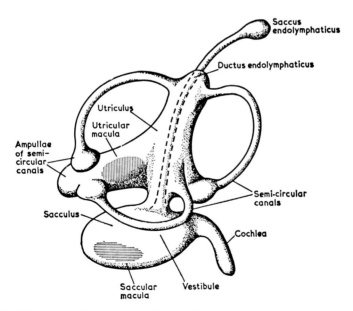

FIG. 12. Diagrammatic representation of the inner ear of a vertebrate. (From Simkiss, 1967.)

within the otic region on either side of the braincase. The epithelial walls of these cavities secrete a fluid, endolymph (see below).

Anatomically, the vertebrate inner ear can be divided into two main compartments, the utriculus and sacculus. Associated with the utriculus are three semi-circular canals. A further canal, the ductus endolymphaticus, originates at the junction of the utriculus and sacculus and terminates in a diverticulum (the saccus endolymphaticus). These structures are depicted diagrammatically in Fig. 12.

ENDOLYMPHATIC SACS

The endolymphatic sacs in most vertebrates are small and contained within the braincase. In the Chondrichthyes, which are presumed to be primitive in this respect, the endolymphatic ducts may extend upwards to open on top of the head (Romer, 1963). Amongst some of the lower vertebrates, for example bony fish, reptiles and especially amphibia, the endolymphatic sac is enlarged and distended with crystals of $CaCO_3$. The anuran frogs possess particularly large endolymphatic sacs which become elongated caudally and penetrate the vertebral canal, often extending backwards along the whole length of the spinal cord (Whiteside, 1922). In these species, the whole system becomes filled with crystals of aragonite, causing the sac to protrude between the intervertebral foramina and giving rise to the typical calcareous (lime) sacs which overlie the spinal ganglia of the adult frog (Simkiss, 1967). The lime sacs are considerably less well developed in Urodela and Apoda (McWhinnie and Cortelyou, 1967, Robertson, 1972b).

Much of the vertebrate inner ear is surrounded by an interstitial fluid, the perilymph, having a composition very like that of the extracellular fluids found in other parts of the body (see Table VIII). Within the sacs and ducts of the inner ear is the endolymph which, in some respects, resembles intersti-

TABLE VIII
Analysis of endolymph and perilymph from some mammals

	Na (mmol litre^{-1})	K (mmol litre^{-1})	Cl (mmol litre^{-1})	Protein (mg %)	Non-protein N_2 (mg %)
Guinea pig					
Perilymph	148	5	120	75	20
Endolymph	26	142	110	25	22
Cat					
Perilymph	164	6	150	142	21
Endolymph	66	117	—	—	—

(After Citron et al., 1956.)

tial fluid, and in others, intracellular fluid (see Table VIII); it is also analogous to bone extracellular fluid (see above). Owing to the relative ease with which endolymph may be collected and analysed, more is known concerning the nature of this fluid. The [potassium/sodium] ratios reported by Johnstone et al. (1963) using microflame spectrophotometry are particularly interesting. These authors reported ratios ranging from 3 in the pigeon (*Columbra livea*) to between 50 and 100 for the guinea pig (*Cavia porcellus*), while ratios in reptiles and amphibia are in the region of 20–40. Clearly the guinea pig ratio is quite different from that reported by Citron et al. (1956) and must reflect the problems associated with obtaining uncontaminated samples from minute areas of tissue and their subsequent analysis.

The electrical potentials across the membrane separating endolymph from perilymph are also rather different from those which might be expected if endolymph were a normal extracellular fluid. They range from less than 13 mV in poikilotherms to between 50 and 100 mV in mammals, while birds show intermediate values of around 20 mV (Schmidt, 1963).

Another characteristic of the endolympatic system is the presence of carbonic anhydrase in the labyrinthine walls. The concentration of this enzyme is variable, but is particularly high in the endolymphatic sac and cochlea, at least in mammals (see Table IX).

All the above characteristics indicate rather specialized metabolic functions for parts of the inner ear, which are not necessarily related to those of hearing and balance.

TABLE IX

The concentration of carbonic anhydrase in the labyrinth of the cat's ear compared with other organs

Organ	Carbonic anhydrase (U gm wet wt^{-1})
Kidney	275
Blood	943
Inner ear—basal turn of cochlea	1086
middle turn of cochlea	2263
apex of cochlea	3162
saccus endolymphaticus	3000
vestibule	81

(After Erulkar and Maren, 1961.)

OTOLITHS

There appear to be two distinct groups of Ca deposits within the inner ear given the general name Otoliths (Simkiss, 1967). These are (1) the statoliths which are associated with the sensations of balance and are common to all vertebrates, and (2) deposits associated with the endolymphatic sacs which are found in only a few vertebrates, notably the anuran amphibians. The endolymphatic $CaCO_3$ deposits appear to be involved in both Ca and acid-base regulation since they represent a major reserve of buffer-base. The structures of vertebrate statoliths are quite variable according to Carlstrom (1963). In cyclostome fish they do not consist of $CaCO_3$, but of apatite. The cyclostomes are unique in this respect. Statoliths in higher fish are of three different polymorphic forms of $CaCO_3$. These are aragonite in Chondrichthyes, vaterite (a rather unstable form) in the Chondrostei and aragonite in Teleostei. The Holostei have statoliths composed of a vaterite-aragonite mixture while the Coelacanthii and Dipnoi have statoliths of aragonite. Amphibian statoliths consist of aragonite while in higher tetrapods they consist of calcite, the most stable form of Ca carbonate. Reptilian statoliths are a mixture of calcite and aragonite. Simkiss (1967) considers that the transition through various polymorphic forms of Ca salts in the vertebrate inner ear represents an evolution to more stable forms.

The amount of calcareous deposit associated with the endolymphatic sacs is very variable. In the cyclostomes the sac is filled with poorly crystalline material. Crystals of $CaCO_3$ are found in the sacs of chondrichthyan fish and teleosts, while in Dipnoi, the sacs appear to be well developed with many diverticuli filled with calcareous material (Simkiss, 1967), a fact to be borne in mind when considering the possible role of endolymphatic lime deposits in the evolution of vertebrates from an aquatic to terrestrial mode of existence (see Chapter 14). While the degree of endolymphatic sac calcification is greatest in amphibia, even in this class some urodelan and apodan species such as *Necturus* and *Siren* may have small sacs with little or no calcareous deposits. Only in anuran frogs are the sacs greatly extended, even penetrating the vertebral column, and contain large quantities of $CaCO_3$ (Simkiss, 1967). In adult reptiles the endolymphatic sacs are small except in gekkonids and iguanids. In these species the sacs extend into the cranium and contain large quantities of $CaCO_3$ deposits (Hamilton, 1964). In birds and mammals the endolymphatic sacs are small and only contain calcareous material during embryonic development (Simkiss, 1967).

Of great interest is the fact that while the Ca deposits of the statoliths do not appear to be metabolically available, those within the endolymphatic sacs are. This is consistent with a role for the statoliths in balance and for endolymphatic $CaCO_3$ as a labile store of Ca and buffer-base.

A further discussion of the metabolic role of amphibian endolymphatic Ca deposits is in Chapter 11.

Dentine

Dentine is a tissue with some similarities to bone and may have evolved from a primitive bony tissue. It is found in the early fossil vertebrates as the outermost layer of the dermal armour which is thrown into a series of ridges or tubercles. Modern dentine is found in the fish scales and teeth of many vertebrate classes. The structures of these different dentines have been surveyed by Halstead (1974). They are remarkably similar in that the dentine is penetrated by a system of branching tubules which radiate out from a central pulp cavity. These tubules appear to be involved in the sensitivity of dentine which most of us have experienced when our tooth enamel erodes to expose the underlying dentine. Halstead (1974) suggests that a function of dentine in fish scales or dermal armour could be as an organ of sensation.

Cells which are associated with formation of dentine are termed "odontoblasts". These retreat as the dentine is formed so that dentine is generally an acellular tissue. In one group of primitive fossil vertebrates, the jointed-necked, armoured Arthrodires which were forerunners of the modern sharks, the odontoblasts became trapped in the matrix of the developing dentine and so became odontocytes. Their lacunae were ovoid and there were no canaliculi.

The essentially acellular structure of mature dentine might suggest that it is not a particularly labile tissue and, therefore, unlikely to play a major role in Ca regulation. However, there is considerable fossil evidence that worn or damaged dentine may be replaced by the activity of odontoblasts which form thin sheets of new dentine overlying the old, while the older underlying tubercles may have been resorbed by odontoclasts (Halstead, 1974).

The matrix of dentine, like that of bone, consists of collagen with some inclusion of proteoglycans. In the portions of tubules near the pulp where there is a highly calcified zone, little collagen is present, but there are large quantities of proteoglycan matrix. This applies to both modern tooth and probably also to the fossil heterostracans (Halstead, 1974).

Dentine is also present in the scales of many extant fish. The Ca reserves within scales are very labile as we shall see later, but it is not clear which fraction of Ca, whether dentine or the osseous portion, is responsible for this.

Other Calcified Tissues in Vertebrates

Bone and dentine which occur mainly as Ca apatites, and endolymphatic lime deposits in which the predominant salt is $CaCO_3$, are the major calcified tissues which contribute to Ca regulation in vertebrates. The important feature of these tissues is their ability to undergo resorption to a greater or lesser degree.

Other vertebrate hard tissues include aspidin, enamel, certain keratinized structures and eggshells of birds and some reptilian species.

Aspidin is thought to be a primitive type of acellular bone, found only in fossil vertebrates beneath the external covering of dentine in the dermal armour. This material, which is related to both bone and dentine, has been described by Halstead (1974). The evolution of aspidin shows a transition from a dentine to a bone-like substance.

Enamel, found as a covering to teeth in higher vertebrates, is a hard, glass-like substance which is protective in function. Fish scales and teeth are covered in a material which resembles enamel and is termed enameloid; however, this is mesodermal in origin while true enamel has an ectodermal origin. In reality, enameloid is an extension of the dentine moiety of a tooth or scale.

Certain keratinized structures contain impregnations of Ca salts. These include hair, reptilian scales, finger nails etc. These are dead tissues which are eventually sloughed off and replaced continuously.

Vertebrates, which lay megalecithal eggs, for example birds and many reptiles, have developed a form of calcified structure, the eggshell, which is composed mainly of $CaCO_3$. This functions as a protective covering for the developing embryo and also provides an important source of Ca for embryonic bones. As an ovum moves along the oviduct, first the albumen is added, followed by the shell membranes. When it reaches the so-called uterus or shell gland, calcification of the eggshell begins. Secretion of Ca and carbonate by the oviduct, a process which has been studied in great detail, will be reviewed later in this book. The process of eggshell secretion is almost certainly regulated by hormones.

Eggshells grow by a process of crystal seeding which is similar to that of other calcified tissues (Halstead, 1974). The initial phase of calcification is the production of spherulites, which are composed of radial spicules of aragonite (a form of $CaCO_3$) which grow around an organic core. The organic material appears to be a proteoglycan (chondroitin sulphate). From the initial spherulites are formed palisade layers of calcite which grow outwards from the shell membranes. Between the calcite layers are pores which allow the embryo to exchange respiratory gases. Following calcification of the eggshell, which in a domestic chicken takes about 18 h, a thin cuticle is deposited over its surface.

The process by which a developing embryo is able to withdraw Ca from the eggshell is under increasing scrutiny (Simkiss, 1975), but it is not within the scope of this book to consider this process in any detail.

Soft Tissues as Calcium Stores

In mammals, calcification of soft tissues occasionally occurs, but this is a pathological situation. There is no evidence that cells can act as specialized tissue reservoirs for Ca since such a function would probably be highly disruptive to normal cellular membrane function and metabolism. However, it is clear that some cell organelles such as mitochondria can accumulate high concentrations of the cation (Glimcher, 1976).

It has been suggested that muscles of teleost fish can act as Ca reservoirs since they contain about five times as much Ca as mammalian muscle (Chan, 1972). The same author found that "removal" of the ultimobranchial gland resulted in a decline in muscle Ca content, while removal of Stannius corpuscles resulted in a rise in this parameter. These changes merely reflect plasma Ca responses to surgery and are not specific hormone responses. Furthermore, it is difficult to see how the muscle cells could act as a major Ca reservoir without serious disruption of the excitation-contraction coupling process in that tissue.

5. Calcium Regulating Hormones

General

We have seen that plasma calcium (Ca) levels throughout the vertebrate Subphylum are regulated to within narrow limits. This is accomplished, in part, by exchange with Ca in the environment and partly by exchange with the hard tissue Ca reservoirs which have been discussed previously. These exchanges are ultimately dependent upon cellular processes which, in turn, are regulated by hormones. In this chapter, the general nature of Ca regulatory hormones is reviewed, while their specific target organs are considered in the following chapter.

Many hormones are recognized as having some effect on Ca homeostasis; these include the three major systems of mammals and other tetrapods, i.e. calcitonin, parathyroid hormone and the active metabolites of vitamin D_3. In addition, hormones such as corticotrophin, prolactin and hypocalcin exert a considerable influence on Ca regulation, at least in the lower vertebrates, while other glands, such as the gonads and adrenal cortex, may also influence Ca metabolism in many, if not all, vertebrate classes.

Parathyroid Hormone

The hypercalcaemic hormone from parathyroid glands was discovered by Collip (1925) and was the first Ca regulating hormone to be recognized. In phylogenic terms, it is the most recent of known Ca regulating hormones to have evolved.

Morphology

For a recent review of the comparative anatomy, histology and embryology of the parathyroid glands, the reader is referred to Roth and Schiller (1976). Parathyroid glands are present in tetrapods but not in fish or primitive vertebrates. In urodelan amphibians the glands do not appear until metamorphosis, that is at the time of disappearance of the gills (Roth and Schiller, 1976). Like the calcitonin secreting ultimobranchial bodies, the parathyroids are apparently derived from the pharyngeal pouches (see Fig. 13). The parathyroids originate as paired structures from the endoderm of the third and fourth pouches, while the ultimobranchials are derived from the sixth pair of pouches.

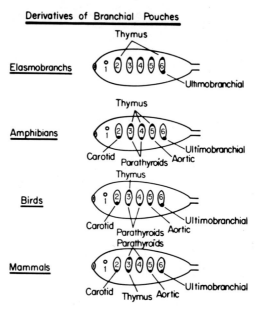

FIG. 13. Diagrammatic representation of the embryonic origin and glandular derivatives of the branchial (gill) pouches in various vertebrate classes. (From Copp, 1969b.)

The embryology of the parathyroid glands is still in question since while they were considered for many years to be of endodermal derivation, recent evidence indicates that they originate from a specialized region of the neuroectoderm (Pearse and Taylor, 1976). This is interesting in view of possible relationships between parathyroid hormone and hormones from the pituitary gland (see below). Embryologically there are three pairs of parathyroids in tetrapods, but those from the second branchial pouches generally disappear. Two or more, but rarely one pair of the glands, persist in adult animals. Amphibians usually have two pairs of glands, while in reptiles, the number may vary between one pair in Crocodilia, two pairs in Chelonia and Ophidia and up to three pairs in Lacertilia. Birds and mammals may have one or two pairs (Bentley, 1976; Roth and Schiller, 1976). There seems, then, to be no particular pattern of evolution regarding the distribution of parathyroids in tetrapods. The glands are usually found in association with the thyroids and ultimobranchials. The arrangement in birds (see Fig. 14) is fairly typical. In mammals, the parathyroids may be embedded in, or at least exist in very close anatomical association with, the thyroid glands. In sub-mammals, the para-

CALCIUM REGULATING HORMONES 43

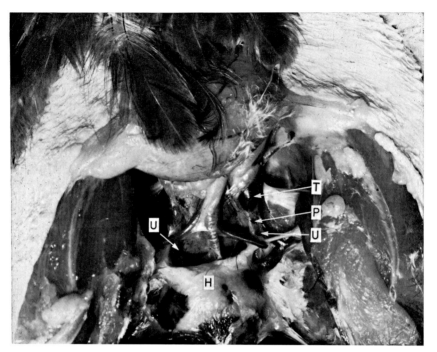

FIG. 14. Dissection of the ventral surface of the neck of the fowl to show the position of the thyroids (T), parathyroids (P) and ultimobranchial glands (U) on the left-hand side of the body. On the right-hand side of the body the ultimobranchial gland is often slightly deeper and more posterior in position. Heart (H). (From Simkiss and Dacke, 1971.)

thyroids tend to be more discrete so that their surgical extirpation is somewhat simpler than in mammals.

Histologically, parathyroids are different from both thyroid and ultimobranchial glands in tetrapods. They do not have a follicular structure but instead comprise whirls of compact epithelial cells. In some vertebrates such as birds, the parathyroids contain clusters of ultimobranchial C cells, while similarly the ultimobranchial bodies may contain nodules of parathyroid tissue. There is considerable variation in the quantity of not only stromal fat, but also of vascularization and fibrous connective tissue. However, the secretory chief cells are essentially similar in all tetrapod classes (Roth and Schiller, 1976).

CHEMISTRY

The clinical importance of parathyroid hormone in maintaining plasma Ca levels has been recognized for more than 50 years, and for this reason, studies of its chemical nature have been largely confined to the hormone from human

44 CALCIUM REGULATION IN SUB-MAMMALIAN VERTEBRATES

or other mammalian sources. Only recently, with the extraction and purification of chicken parathyroid hormone by MacGregor *et al.* (1973), has the hormone from sub-mammalian sources begun to be studied.

Parathyroid hormone is a polypeptide consisting of 84 amino acids. Amino acid sequences for porcine, bovine and the amino terminal portion of the chicken molecule are depicted in Fig. 15. The complete amino acid sequences

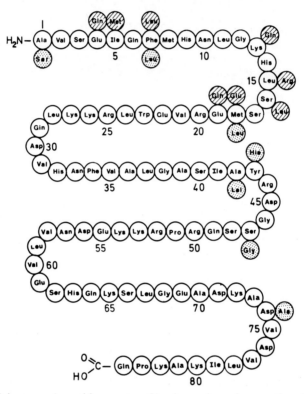

FIG. 15. Primary amino acid sequence of bovine and porcine parathyroid hormones. The backbone sequence is from the bovine hormone. Dotted area indicates porcine substitutions along complete molecule. Hatched area indicates chicken substitutions to position 20. (From Parsons, 1976, with additional data from MacGregor *et al.*, 1973.)

for parathyroid hormones from the sub-mammals are, as yet, unknown. The first of these to be elucidated will be the chicken hormone since it has already been extracted and purified to a high degree and the amino-terminal sequence to position 20 is now known (see Fig. 15).

Very recently, Snell and Smyth (cited by Parsons, 1976), have suggested a structural homology between parathyroid hormone and corticotrophin. This is illustrated in Fig. 16, in which the sequence of bovine parathyroid hormone between residues 15 and 25 and corticotrophin residues 1 to 11 are compared. It is of interest that corticotrophin residues 4–10 constitute the heptapeptide core which is common to corticotrophin, β lipotrophic hormone, and melanocyte stimulating hormone. This core is essential for the biological activity of these hormones (Parsons, 1976). Such a homology is extremely striking when its statistical chances are considered, particularly as the tryptophan and methionine residues common to both parathyroid hormone and the pituitary hormones are found rather infrequently. There is, then, a possibility that the tetrapod parathyroid hormone may have evolved from a pituitary hormone in ancestral fish. Certainly the pituitary in fish, as in tetrapods, appears to produce hypercalcaemic hormones, the nature of which will be considered below.

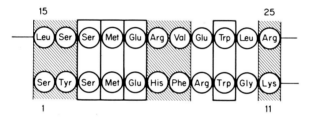

FIG. 16. Comparison of a portion of the sequence of the biologically active region of bovine parathyroid hormone (bPTH 15–25) with the amino-terminal 11 residues of corticotrophin from the same species. This contains the heptapeptide core region (ACTH 4–10) which is common to corticotrophin, lipotropin and the melanocyte-stimulating hormones and is essential for their biological activity. Identical residues are enclosed in solid lines; residues related by the criteria of Barker and Dayhof (1972) are joined by hatching. If position 22 in bPTH is reassigned as glutamine it becomes possible to extend the hatched area, since Glu can arise from Arg by a single mutation in the codon. (From Parsons, 1976.)

Bovine and porcine parathyroid hormones exhibit similar hypercalcaemic activity in the parathyroidectomized rat bioassay, but the human hormone is only about one-third as potent. The complete molecule is not essential for biological activity since the 1–34 peptide has similar, if not slightly greater activity than the complete 1–84 molecule (Tregear et al., 1973), indicating that considerable polymorphism of the molecule may occur. Native bovine parathyroid hormone appears to consist of three isohormones (Keutmann et al., 1971), of which the major type, bovine parathyroid hormone I, is depicted in Fig. 15.

It seems surprising that until recently no reports had appeared concerning attempts to extract and purify parathyroid hormones from sub-mammalian classes of vertebrates, especially since the glands are often discrete in these classes and, therefore, suitable for harvesting. Presumably a number of attempts have been made to extract these sub-mammalian hormones. Dacke (unpublished observations) extracted glands from Japanese quail (*Coturnix*

coturnix japonica) by conventional techniques and tested them for hypercalcaemic activity by the sensitive avian bioassay method of Dacke and Kenny (1973); however, no activity could be found. The secret of extracting parathyroid hormone from avian sources seems to have been discovered by MacGregor *et al.* (1973). This group were able to stimulate the parathyroid glands of immature chickens by feeding diets deficient in vitamin D_3. The hyperplastic glands were subsequently extracted and found to yield assayable quantities of parathyroid hormone. Chemically, this material is similar in molecular size and charge to its bovine counterpart. Its activity in a mouse calvarium assay system revealed a potency of 8000 U.S.P. U mg^{-1}. It also cross-reacted, albeit weakly, with three different antisera to bovine parathyroid hormone.

Pro-parathyroid Hormone

Many peptide hormones, including those from the parathyroid and ultimobranchial glands, contain relatively few amino acids and are formed as a result of disassembly from larger protein or polypeptide molecules. Thus the formation of polypeptide hormones often involves the initial formation of an inactive parent molecule called a "pro-hormone". This mechanism prevents the initial genetic translation via messenger RNA into a hormone which could be active within its tissue of origin. As a result of post-translational changes, such as enzymic cleavage, pro-hormone is reduced to the active hormone which is released into the circulation.

The presence of a pro-parathyroid hormone in bovine parathyroid glands was first reported by Hamilton *et al.* (1971). This molecule is larger than the 1–84 native parathyroid hormone; it consists of 109 amino acids and has a mol. wt of 12 000 compared with 9563 for 1–84 peptide (Cohn *et al.*, 1972). Similarly, chicken parathyroid glands appear to contain a pro-hormone having similar properties to its bovine counterpart (MacGregor *et al.*, 1973).

An even larger chain to be initially synthesized is pre-pro-parathyroid hormone consisting of 115 amino acids. Like pro-hormone, the further extension in the pre-pro-hormone is at the amino-terminus and this undergoes cleavage to form pro-parathyroid hormone before being released from the endoplasmic reticulum (Habener *et al.*, 1976).

Assay Methods

It is not within the scope of this book to consider in detail the various assay methods for parathyroid hormone. For a recent review of this subject the reader is referred to Kenny and Dacke (1975).

In this section, assay systems involving the use of whole animals or their tissues, from sub-mammalian classes, will be considered, as well as the use of mammalian systems for assay of sub-mammalian parathyroid hormones. Hormone assay systems can be subdivided into three major types. These include *in vivo* bioassays, *in vitro* bioassays and radio-immunoassays. All these methods have advantages and disadvantages. *In vivo* bioassays are often quick and simple, but lack specificity and precision. Radio-immunoassays are precise and highly specific, but their high specificity can be a disadvantage. For instance, it is not practical to assay sub-mammalian parathyroid hormone using mammalian antisera. The low cross-reactivity of chicken parathyroid hormone with bovine antisera has been referred to in the preceding section (MacGregor *et al.*, 1973). In order to assay low levels of hormone, for example in the circulation, radio-immunoassays are generally used. In the case of chicken parathyroid hormone, obviously an antiserum to the chicken hormone requires to be developed before such an assay becomes feasible. *In vitro* bioassays combine some advantages of both *in vivo* and radio-immunoassay systems. They are relatively sensitive and specific, but may also be tedious in preparation and slow in response time. Recently, *in vivo* assay systems to parathyroid hormone involving birds have been developed. These assays were developed independently in three separate laboratories (Dacke and Kenny, 1971; 1973; Lewis and Taylor, 1972; Parsons *et al.*, 1973). The method is simple, economical and reasonably sensitive and precise and represents a considerable improvement over similar assays involving mammals. Polin *et al.* (1957) were the first to emphasize the rapidity (3–4 h) of the hypercalcaemic response to parathyroid hormone in 5–7-week-old chickens and to suggest this response as a possible basis for a bioassay. However, their method lacked sufficient sensitivity and precision to become widely used. Dacke and Kenny (1971) reported that 2–3-week-old Japanese quail or chickens respond to parathyroid hormone with a hypercalcaemic response which is both rapid (30–60 mins) and sensitive (> 2 U.S.P. U per bird), although the precision of this response was too poor ($\lambda = > 0.4$)* for the method to be used as a bioassay. This advance led to the successful development of an assay method based on the hypercalcaemic responses of 2–3-week-old Japanese quail or 5–6-day-old chickens (Dacke and Kenny, 1973). The assay is rapid (60 mins), sensitive (0.4–1.2 U.S.P. U per bird) and reasonably precise ($\lambda = 0.20$).* In order to enhance sensitivity and precision in this assay it was found necessary to incorporate CaCl (51 mmol litre^{-1}) into the injection media. The chicken bioassay method of Parsons *et al.* (1973) is remarkably similar in that CaCl is also added at a dose level of 20 mmol per chick (approx. 50 mmol litre^{-1} assuming an injection volume of 0.4 ml). Lewis and Taylor (1972) did not incorporate CaCl into their injection media when assaying parathyroid hormone in one-day-old chicks. Apart from the relative sensitivity of avian bioassays to mammalian parathyroid hormone, they have an added advantage that no prior parathyroidectomy is necessary in contrast to most bioassays involving rodents.

The quail bioassay has now been successfully applied to the detection (using

* The index of precision (λ) is obtained by dividing the s.d. of the assay by its slope. The slope (b) is the difference between the mean responses to the high and low doses of hormone divided by the logarithm of the dose interval. The lower the value of λ, the more precise is the assay.

Sephadex G50 filtration for recovery and concentration) of parathyroid hormone in small aliquots (< 50 ml) of urine from human patients with chronic renal failure (Kenny et al., 1976).

The mechanism of the hypercalcaemic responses of birds to parathyroid hormone is interesting and is discussed in Chapter 12.

CONTROL OF SECRETION

Of the major Ca regulating hormones in vertebrates, control of parathyroid hormone secretion seems to be the least complex. Perhaps this reflects the relatively late phylogenic appearance (in amphibia) of this hormone in contrast to the appearance of calcitonin and vitamin D_3 which first occurred in more primitive acquatic vertebrates. Mechanisms of parathyroid hormone release have been studied extensively in mammals by virtue of the introduction of sensitive radio-immunoassay methods, enabling circulating levels of hormone to be detected, but not so in sub-mammals. Until proven otherwise, we must assume that general mechanisms of release for the hormone also hold true for the lower orders of tetrapods.

In common with many other hormones, secretion of parathyroid hormone is under the regulation of a short loop negative feedback system. Thus when the level of plasma Ca falls, secretion of hormone is stimulated. Alternatively, when plasma Ca levels rise, hormone secretion is inhibited (Copp, 1968). A similar inverse relationship exists between plasma magnesium levels and parathyroid hormone secretion, at least with high concentrations of the magnesium ion (above 3 mmol litre^{-1}), when it is as potent as the Ca ion as a regulator of parathyroid hormone synthesis and secretion (Sherwood et al., 1972).

Recently, 3′, 5′-cyclic AMP has been implicated as a mediator of the cation effects on parathyroid hormone release (Dufresne and Gitelman, 1972; Sherwood et al., 1972). These groups demonstrated that addition of dibutyrylcyclic AMP or theophylline to explanted parathyroid tissue *in vitro* resulted in increased levels of parathyroid hormone secretion. It was suggested that the hormone release might be mediated by a magnesium dependent adenyl cyclase system with low concentrations of magnesium ion inhibiting rather than stimulating release of the hormone. Thus 3′, 5′-cyclic AMP appears to mediate release of parathyroid hormone as it does for several other endocrine systems such as the thyroid, adrenal cortex, ovary and possibly the adenohypophysis and pancreatic β cells (Bentley, 1976). Apart from the Ca and magnesium ions, there is some evidence that the acid-base status of the blood may also have a controlling influence on parathyroid hormone secretion, although the situation appears to be rather complex. When dogs were rendered acidotic or alkalotic by infusion of hydrochloric acid or sodium bicarbonate solutions, recovery from an oxalate induced hypocalcaemia was markedly retarded in acidotic dogs, but alkalotic dogs recovered faster than neutral controls. Alkalosis, however, failed to increase slow or non-recovery in previously thyroparathyroidectomized dogs (Fujita et al., 1965). It was suggested that this response was related to parathyroid secretion which was presumably blocked by acidosis and enhanced by alkalosis. Toribara et al. (1957) demonstrated that pH has a marked effect on the distribution of bound and ultrafilterable Ca in the plasma *in vitro*. Thus, the higher the pH of plasma, the lower is the proportion of ultrafilterable to bound Ca, while lowered pH has the opposite effect. This being the case, it can be argued that a raised pH will indirectly stimulate release of parathyroid hormone by lowering the ultrafilterable and hence ionized fraction of Ca. Certainly the changes in ultrafilterable Ca fractions noted by Toribara et al. (1957) would be sufficient to account for a fairly large change in the rate of hormone secretion. Presumably this system in reverse would function with respect to calcitonin secretion. *In vivo*, however, the situation might be different since there is evidence that metabolic acidosis may actually lower the ionized Ca level in plasma by promoting hyperphosphataemia which indirectly can cause an increase in parathyroid hormone secretion (Prien et al., 1976).

Humoural agents such as glucagon, gastric hormones and gonadal hormones, all of which are thought to play a role in calcitonin secretion (see below), have not been shown to mediate secretion of parathyroid hormone. There are, however, some reports that calcitonin itself may act as a secretagogue for parathyroid hormone, at least *in vitro* at high doses (100 ng ml^{-1}). It is suggested that calcitonin might activate the adenyl cyclase system in the parathyroid secreting cells as it does in bone and kidney (Habener et al., 1976). There is some evidence that the vitamin D_3 metabolite, 24, 25-dihydroxycholecalciferol, when perfused through isolated goat parathyroid glands *in vivo*, may inhibit secretion of parathyroid hormone (Bates et al., 1975). Thus, a concentration of the metabolite at 1 ng ml^{-1} in the perfusion medium reduced parathyroid hormone secretion to about 40% of the control value. The metabolite 25-hydroxycholecalciferol and its synthetic analogue 1 a-hydroxycholecalciferol at a dosage of 100 ng ml^{-1} and 10 ng ml^{-1} respectively, were without effect in this system. Other experiments with 1,25-dihydroxycholecalciferol indicate that at high concentration (25 ng ml^{-1}) this substance is also without effect on parathyroid hormone secretion, but paradoxically, a lower dose (125 pg ml^{-1}) caused a significant reduction in secretion of the hormone (Brumbaugh et al., 1974). There are, as yet, no reports of any neural or other higher influences on parathyroid hormone secretion.

Glandular levels of parathyroid hormone are generally much lower than those of other hormones, for instance, in the pituitary gland. This is probably due to a high rate of intracellular degradation (Parsons, 1976). It was suggested by Potts (1976) that higher rates of secretion might be achieved by a reduction of intracellular degradation as well as by an increase in the rate of biosynthesis.

CIRCULATING PARATHYROID HORMONE

Circulating parathyroid hormone may exist as several molecular species or fragments ranging in mol. wt from 4500–9500 (Kenny and Dacke, 1975;

50 CALCIUM REGULATION IN SUB-MAMMALIAN VERTEBRATES

Parsons, 1976). Both the intact molecule (mol. wt 9500) and the smaller fragments may have immunological as well as biological activity. By using parathyroid hormone labelled with ^{125}I or ^{131}I it has been possible to study metabolic degradation of the hormone *in vivo*. The half-life of biologically active parathyroid hormone in mammalian blood varies from 22 min in rats (Malick *et al.*, 1965) to 19–20 min in cows (Sherwood *et al.*, 1966). Some caution must be exercised, however, when considering such data since Yalow and Berson (1966) have demonstrated that ^{131}I labelled bovine parathyroid hormone is degraded at a slower rate *in vivo* than unlabelled hormone. More recent studies from Blum *et al.* (1974) in which Ca was infused in order to arrest endogenous parathyroid secretion in unanaesthetized cows, suggest that native 1–84 peptide may have a half-life as low as 2–4 min. It seems, however, that a large proportion of the cleavage products still retain biological activity before being further degraded to inactive fragments in the kidney and liver (Parsons, 1976).

At this point it is worth speculating about the rate of parathyroid hormone degradation in the blood of sub-mammalian species (and this can only be speculation). It is generally accepted that the hypercalcaemic responses to injected bovine parathyroid hormone are more rapid in sub-mammals compared with those in mammals. For example, frogs exhibit a hypercalcaemic response to 10 U.S.P. U of parathyroid hormone within 2 h (Cortelyou, 1967), while in birds, the responses are even faster, 10 min in chickens (Candlish and Taylor, 1970) and 15 min in Japanese quail (Kenny and Dacke, 1974). When comparing these responses with those of mammals it is important to avoid the common

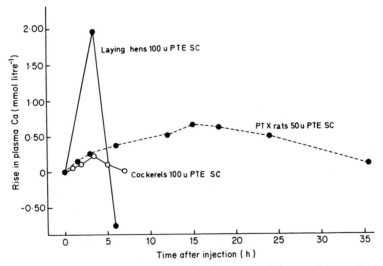

FIG. 17. Effect of subcutaneous parathyroid extract injection in the fowl and parathyroidectomized rat. Note the marked difference in sensitivity of the egg-laying hen from the cockerel. The parathyroidectomized rat takes longer to achieve a peak hypercalcaemic response but initiation of this response is as rapid as the fowl. (Data from Polin *et al.*, 1957 and Rasmussen and Westall, 1956.)

misconception that mammalian responses to parathyroid hormone are slow; they are not. Careful examination of the hypercalcaemic response in rats (Fig. 17) reveals that it is apparent within an hour or so of subcutaneous injection. The major differences between mammalian responses and those of sub-mammals, at least birds, are their sensitivity and transience compared with those in mammals. The avian response to 100 U.S.P. U parathyroid hormone has disappeared within 6 h, while that in rats is only reaching its peak by 15 h. Birds, however, appear to show a response of greater amplitude than that in mammals. This general comparison holds true for most avian and mammalian hypercalcaemic responses to the hormone. Two main conclusions may be drawn from this data. Firstly, it would indicate that target organ receptors in birds are more sensitive to mammalian parathyroid hormone than are mammalian ones. This goes against the general trend for vertebrate hormones which tend to be less effective in heterogenous species or classes than in the homologous species from which they originate (Bentley, 1976). Secondly, it would suggest that parathyroid hormone of mammalian origin is degraded more rapidly in birds than in mammals. It would be surprising if avian studies with labelled hormone do not reveal much briefer half-lives in the circulation than the 20 min or so common in mammals for biologically active hormone.

Parathyroid hormone levels in mammalian circulation have been reported using both radio-immunoassay and more rarely, sensitive bioassay techniques (Kenny and Dacke, 1975). In general, it would seem that the hormone levels reported by bioassay are about hundredfold greater than those reported using radio-immunoassay, thus demonstrating the difficulties involved in attempting to compare assay techniques. The normal secretion rate of hormone in dogs and cows is about 0·1 U.S.P. U [50 ng (kg body wt)$^{-1}$ h^{-1}]. This would give a plasma level of less than 10 pg ml^{-1} for the intact 1–84 peptide, a level at which it is very difficult to detect the hormone in peripheral blood by conventional N-terminal radio-immunoassays (Parsons, 1976). At the time of writing this book, levels of circulating parathyroid hormone in sub-mammalian species have not been reported.

Calcitonin

Calcitonin was discovered relatively recently, its existence first being proposed by Copp et al. (1962). The controversy surrounding the initial discovery of this hormone, and its subsequent settlement, is a classical example of the value of comparative research. The studies of Copp et al. involved experimental perfusion of the thyroid-parathyroid complex in dogs. Blood containing artificially raised or lowered Ca concentrations was perfused into the arteries of these tissues and the resulting venous outflow redirected into the general circulation of the dog. Effects of this treatment on general systemic, plasma Ca levels were observed and it was found that when the perfusate contained low concentrations of Ca, a hypercalcaemic response occurred consistent with an increase in parathyroid hormone secretion. When the perfusate contained high concentrations of Ca, a hypocalcaemic response occurred. This latter response was rapid, in fact much faster and bigger than might be expected if it were due merely to a decline in the level of circulating parathyroid hormone which would occur, for example, if the thyroid-parathyroid complex had been completely removed. This led to the proposal by Copp et al. (1962) for a hypocalcaemic factor, "calcitonin".

Meanwhile Hirsch et al. (1963) had noticed that during routine parathyroidectomy of rats, prior to their use for parathyroid hormone bioassay, the fall in plasma Ca levels following surgery was greater if the glands were destroyed by electrocautery rather than by simple surgical excision. This observation led them to suspect the presence of a hypocalcaemic factor which was subsequently extracted from the rat thyroid, purified and named "thyrocalcitonin". The initial controversy concerning the parathyroid versus thyroid origin of the hypocalcaemic hormone was eventually clarified by the elegant studies of Copp et al. (1967a) who made extracts from the ultimobranchial bodies of sub-mammalian vertebrates and injected them into young rats to produce hypocalcaemia. The tissue was found to constitute a rich source of the hypocalcaemic hormone calcitonin. In sub-mammals, the ultimobranchial bodies remain more or less discrete, but in mammals they become incorporated into the thyroid to form thyrocalcitonin secreting C cells. C cells may also be found in small quantities in the parathyroid tissue and other areas of the neck. Thus calcitonin and thyrocalcitonin were demonstrated as being one and the same hormone.

ULTIMOBRANCHIAL MORPHOLOGY

Calcitonin secreting C cells are present in the mammalian thyroid, but rarely in that of sub-mammals (Pearse, 1976). Embryologically they are derived from ultimobranchial bodies which are present in all extant vertebrate classes except for cyclostomes (see Fig. 13). Like the parathyroid glands, it seems that the ultimate embryonic source of C cells lies in the neuroectoderm of the neural crest. Thus C cells, in common with other polypeptide secreting cells of neuroectodermal origin, possess the ability to take up and decarboxylate amine precursors, the so-called APUD (amine precursor uptake and decarboxylation) phenomenon (Pearse, 1976).

Whereas parathyroid glands lack a follicular structure, the ultimobranchial bodies do not; instead, their structure may vary from a simple gland of cylindrical appearance consisting of two or three layers of granular secretory cells surrounding a single central antrum in some fish such as the eel (*Anguilla anguilla*), to a more complex multifollicular structure as in birds. The main characteristics of C cells are the poorly developed, rough, endoplasmic reticulum and the specific secretory granules which are surrounded by a single membrane. The cells have abundant mitochondria, a well-developed

Golgi region and large numbers of free ribosomes (Pearse, 1976). In the same review the comparative morphology of vertebrate ultimobranchial bodies and the cytochemistry of these tissues are also considered. Studies in a variety of hibernating species of mammals indicate the existence of an annual C cell cycle. Thus in spring, the C cells of bats, hedgehogs, marmots and dormice are well granulated like those of non-hibernators. High synthetic activity in the autumn is followed by degranulation and subsequent exhaustion of the cells (Pearse, 1976). However, it appears that these cycles are more related to season than hibernation *per se* since in November and December, thyroidal calcitonin contents fall markedly in both normothermic and hibernating ground squirrels (*Citellus tridecembineatus*) (Kenny and Musacchia, 1976). Although the ultimobranchials are generally separate from the thyroid in sub-mammals, in some species there may be a degree of association. Thus in chickens the ultimobranchials may contain parathyroid tissue, while in pigeons the thyroid contains appreciable amounts of C cells (Copp, 1972; Pearse, 1976). The admixture of the three endocrine tissues in vertebrates, the thyroids, parathyroids and ultimobranchials, may be significant. While the nature of this relationship is not known, it has been suggested that the association of these tissues may have some functional relevance (Bentley, 1976).

Since both modern chondrichthyan and osteichthyan fish possess ultimobranchial tissue and calcitonin, it is clear from reference to Fig. 1 (see Chapter 1) that the hormone must have appeared at an early stage in vertebrate evolution. The early ostracoderms and placoderms were bony creatures as, presumably, were the main lines of ancestral vertebrates, and it seems likely that evolution of the hormone, calcitonin, was approximately concurrent with that of bone. This must raise the question of whether or not the primary function of calcitonin was associated with the maintenance of a labile bone based on salts of Ca and phosphate. We shall return to this problem later.

Another question concerns the role of calcitonin in modern chondrichthyans (sharks and related species). These fish have cartilaginous skeletons, but they undoubtedly evolved from bony ancestors. Does calcitonin have a function in these species or is it merely vestigial?

CHEMISTRY

The hormone has been extracted from ultimobranchial tissues in all extant classes of vertebrates except for the jawless Cyclostomata. An attempt by Copp *et al.* (1970) to prepare extracts from the branchial regions of the hagfish (*Polistrotremata stoutii*) did not yield any material with hypocalcaemic activity in the rat bioassay, leading them to the conclusion that phylogenic development of calcitonin and the ultimobranchial body were coincidental.

Like parathyroid hormone, calcitonin is a polypeptide, but rather smaller in molecular size, consisting of a chain of 32 amino acids (mol. wt 3000) and with a seven membered ring at the N-terminus. Unlike parathyroid hormone, the entire chain of 32 amino acids is necessary for the expression of biological

54 CALCIUM REGULATION IN SUB-MAMMALIAN VERTEBRATES

activity (Potts and Aurbach, 1976). In contrast to parathyroid hormone, calcitonin is very easy to extract in assayable quantities from sub-mammalian tissues. The quantities of calcitonin extracted from glands of various vertebrates are shown in Table X.

Amino acid sequences for calcitonins from several sources have been elucidated, and, as might be expected, these show some variation. In Fig. 18 the structure of salmon calcitonin I is shown, while Table XI compares amino

TABLE X
Calcitonin concentration in glands from various vertebrates

Class and species	Thyroid	U g^{-1}(MRC) fresh gland wt		Unit kg^{-1} body wt
		Ultimo-branchial	Parathyroid	
Mammalia				
Man (*Homo sapiens*)				
Normal thyroid	0·4	—	0·1–0·5	0·16
Medullary cell carcinoma of thyroid	17	—	—	—
Rat (*Rattus rattus*)	5–15	—	—	0·2–0·6
Hog (*Sus scrofa*)	2–5	—	—	0·4–0·8
Dog (*Canis familiaris*)	1–4	—	1·5–3·3	0·25–0·50
Rabbit (*Oryctolagus cuniculus*)				
Lower pole	1·5–2	—	2·1–2·5	—
Upper pole	a	—	—	—
Aves				
Domestic fowl (*Gallus domesticus*)	a	30–120	—	0·5–0·8
Turkey (*Meleagris gallopavo*)		60–100	—	0·5–0·9
Reptilia				
Turtle (*Pseudemys concinna suwaniensis*)	a	3–9	—	0·002–0·006
Amphibia				
Bullfrog (*Rana catesbeiana*)	—	0·5–0·8	—	0·001–0·002
Teleosti				
Chum salmon (*Oncorhynchus keta*)	—	25–40	—	0·4–0·6
Grey cod (*Gadus macrocephalus*)	—	10–20	—	0·2–0·4
Elasmobranchii				
Dogfish shark (*Squalus suckleyi*)	a	25–35	—	0·25–0·40

[a] No detectable hypocalcaemic activity.

TABLE XI

Amino acid sequences of the calcitonins[a]

Species		2		4		6		8		10		12		14		16
Eel	—	Ser	—	—	—	—	—	Val	—	—	Lys	Leu	Ser	—	Glu	Leu
Salmon	—	Ser	—	—	—	—	—	Val	—	—	Lys	Leu	Ser	—	Glu	Leu
Salmon II	—	Ser	—	—	—	—	—	Val	—	—	Lys	Leu	Ser	—	—	Leu
Salmon III	—	Ser	—	—	—	—	—	Met	—	—	Lys	Leu	Ser	—	—	Leu
Human	Cys	Gly	Asn	Leu	Ser	Thr	Cys	Met	Leu	Gly	Thr	Tyr	Thr	Gln	Asp	Phe
Porcine	—	Ser	—	—	—	—	—	Val	—	Ser	Ala	—	Trp	Arg	Asn	Leu
Bovine	—	Ser	—	—	—	—	—	Val	—	Ser	Ala	—	Trp	Lys	—	Leu
Ovine	—	Ser	—	—	—	—	—	Val	—	—	Ala	—	Trp	Lys	—	Leu
		18		20		22		24		26		28		30		32
Eel	His	—	Leu	Gln	—	Tyr	—	Arg	—	Asp	Val	—	Ala	—	Thr	—
Salmon I	His	—	Leu	Gln	—	Tyr	—	Arg	—	Asn	Thr	—	Ser	—	Thr	—
Salmon II	His	—	Leu	Gln	—	—	—	Arg	—	Asn	Thr	—	Ala	—	Val	—
Salmon III	His	—	Leu	Gln	—	—	—	Arg	—	Asn	Thr	—	Ala	—	Val	—
Human	Asn	Lys	Phe	His	Thr	Phe	Pro	Gln	Thr	Ala	Ile	Gly	Val	Gly	Ala	Pro-NH_2
Porcine	—	Asn	—	—	Arg	—	Ser	Gly	Met	Gly	Phe	—	Pro	Glu	Thr	—
Bovine	—	Asn	Tyr	—	Arg	—	Ser	Gly	Met	Gly	Phe	—	Pro	Glu	Thr	—
Ovine	—	Asn	Tyr	—	Arg	Tyr	Ser	Gly	Met	Gly	Phe	—	Pro	Glu	Thr	—

[a] The entire sequence is shown for human calcitonin. Dashes (—) indicate residues identical to those in human molecule. Residues shown for other calcitonins are only those that differ from human calcitonin. (From Potts and Aurbach, 1976.)

56 CALCIUM REGULATION IN SUB-MAMMALIAN VERTEBRATES

acid sequences for several calcitonins. It is obvious that major differences exist between amino acid compositions of calcitonin from diverse species (see Table XI). Even within species several forms of the hormone may exist; there are three different forms, for instance, in salmon, designated as salmon I, II and III (Niall et al., 1969). The major form calcitonin I is represented in Fig. 18. However, multiple forms have not been demonstrated in any species other than salmon at the present time. Amino acid sequences for eel and salmon calcitonins are, not surprisingly, more similar to each other than either of them

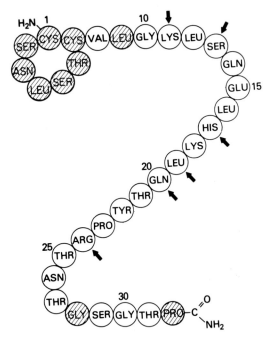

FIG. 18. Structural features of the calcitonin congeners. The molecule shown is salmon calcitonin 1. The hatched residues are common to all 8 molecules analysed to date. The arrows indicate residues common to the piscine calcitonins, the most highly active congeners known. (From Potts and Aurbach, 1976.)

is to the porcine molecule. At the time of writing this book, the amino acid sequence for chicken calcitonin has not been completely elaborated, although the quantitative amino acid composition is known. This is quite similar to that for the eel or salmon hormones. Furthermore, in radio-immunoassay, the chicken hormone gives a completely parallel curve of displacement from that of salmon, reflecting a high degree of immunochemical similarity (Cutler et al., 1974).

Structural differences in calcitonins from different species result in considerable differences in biological activity. Fish (salmon and eel) calcitonins are much more active when tested in mammalian assay systems than are the natural (homologous) hormones. Thus salmon calcitonin I and II and eel calcitonin tested in young rats *in vivo*, assay at 2500–3500 MRC U mg^{-1} compared with a range of 50–200 MRC U mg^{-1} for porcine, ovine, bovine or human hormone. This may be a unique situation in comparative endocrinology and is due to two factors: (1) a slow rate of destruction of the fish hormones in plasma *in vivo* and (2) a greater affinity for the target organ receptors (Marx *et al.*, 1972). This interesting situation suggests that a species need not necessarily have evolved a hormone of such chemical structure to impart maximum biological activity within that species. The chemical evolution of the calcitonins in terms of gene duplication has been discussed by Potts and Aurbach (1976), but the functional significance of this evolution is not as yet known.

While there is some suggestion that parathyroid hormone with its supposedly neuroectodermal origins may have a structural homology with corticotrophin, there is no report of any similar homology between calcitonin and other hormones of neuroectodermal origin.

PRO-CALCITONIN

Evidence for a large mol. wt precursor to calcitonin was first produced by Roos *et al.* (1974). They used a tissue culture method based on trout ultimobranchial cells. When incubated with [^{14}C]-leucine, the cells produced material with immunological similarity to salmon calcitonin. Gel chromatography of this material revealed two distinct peaks of radio-immunoassayable and immuno-precipitable calcitonin-like activity. One peak coeluted with radio-iodinated calcitonin at a mol. wt around 3000, while the other was a high mol. wt (7000) species.

ASSAY METHODS

As with parathyroid hormone, it is beyond the scope of this book to consider assay methods for calcitonin in any detail. All assays currently in use are based on either mammalian *in vivo* or *in vitro* responses to the hormone, or on antisera raised in mammals. For a review of these assay methods the reader is referred to Gray *et al.* (1974) and Munson (1976). Assays involving whole animals or their tissues from sub-mammalian vertebrate species have not been developed. This may be, in part, due to the simplicity and sensitivity of mammalian assays to calcitonin, but is, perhaps, mainly a reflection of the remarkable refractoriness of sub-mammals to this hormone, at least with respect to any hypocalcaemic responses.

CONTROL OF SECRETION

Like that for parathyroid hormone, the main regulatory process affecting calcitonin secretion involves a simple short loop negative feedback, with plasma Ca ion levels as the modulating influence. Thus when the plasma Ca level rises, calcitonin synthesis and secretion are stimulated; when the level of plasma Ca falls, calcitonin synthesis and secretion are inhibited.

In common with parathyroid hormone, there appears to be a relationship between intracellular $3',5'$-cyclic AMP concentration within the thyroid gland and calcitonin release (Care et al., 1970), while infusion with a $3',5'$-cyclic AMP analogue, dibutyryl cyclic AMP stimulates release of the hormone from the thyroid (Bell, 1970).

So far, then, the regulation of calcitonin synthesis and secretion would appear to more or less mirror that of parathyroid hormone. However, regulation of calcitonin is more complex than this. Care et al. (1969) suggested that another hormone, glucagon, when perfused through the thyroid-parathyroid complex in mammals, could effect the rapid release of calcitonin. Later, other hormones derived from the gut, namely pancreozymin-cholecystokinin (Care et al., 1971) and gastrin and its homologues (Cooper et al., 1971) were also demonstrated as potent secretagogues for calcitonin.

It seems likely that the response in plasma calcitonin to a meal is not primarily due to an increase in the plasma Ca level but rather to the release of humoural substances such as gastrin, following stimulation of the gut by the ingestion of food.

In contrast to the situation with parathyroid hormone, both 24,25-dihydroxycholecalciferol (1 ng ml^{-1}) and 1,25-dihydroxycholecalciferol (0·45–3·1 ng ml^{-1}) have no significant effect on calcitonin release in perfused pig thyroids *in vivo* (Care et al., 1975).

There also appears to be a complex relationship between gonadal hormones and plasma or thyroidal calcitonin levels in vertebrates. This was first suggested by Phillipo et al. (1971) who treated mature ewes with daily intramuscular injections of 20 mg progesterone for 7 days. This treatment resulted in a significant mean reduction of 62% in thyroidal calcitonin concentrations when compared with controls. Similarly, calcitonin release from either perfused glands *in vivo*, or slices *in vitro*, was greatly reduced in progesterone treated animals. Other workers have shown indirectly that gonadal function and plasma calcitonin levels are related. Thus rutting stags (*Cervus elephus*) show seasonal elevations in thyroidal calcitonin levels (Phillipo et al., 1971). Furthermore, the mean plasma calcitonin concentration in bulls is significantly higher than in cows (303 pg ml^{-1} vs 165 pg ml^{-1}) according to Deftos et al. (1972a), while in Japanese quail, plasma levels of the hormone are three times higher in the adult male than in laying hens (Boelkins and Kenny, 1973). In salmon (*Oncorhynchus nerka*) however, it is the female which exhibits higher plasma calcitonin levels compared with males (Deftos et al., 1972b). These relationships are complex and, as yet, incompletely understood. It does seem that in a wide range of vertebrate classes, gonadal function can influence plasma calcitonin levels and hence Ca metabolism. This is not surprising in view of stresses imposed on Ca metabolism during reproduction (Simkiss, 1967). These relationships are considered in detail in later chapters.

CIRCULATING CALCITONIN

Levels of calcitonin in the circulations of mammals are not usually detectable by bioassay. However, radio-immunoassay has proved more useful, with the low levels of the hormone reported in human plasma, for instance (50–150 pg ml^{-1}) by Huefner and Hesch (1971). Radio-immunoassays and bioassays for calcitonin give reasonably close agreement in hypercalcaemic rats which have high levels of circulating calcitonin; the value reported by radio-immunoassay was 0·0003 MRC U ml^{-1}, while that reported using an *in vitro* bone culture bioassay was 0·0005–0·001 MRC U ml^{-1} (Gray *et al.*, 1974).

Studies concerned with the degradation of calcitonin in the circulation suggest that heterologous hormones may be much longer lived in certain species than the homologous ones. Thus the half-life for porcine calcitonin in rat plasma *in vivo* is around 2 min, but those for salmon or chicken hormone are 10–20 min (see Fig. 19). Eel calcitonin may be even longer lived, at least in human serum, where it is inactivated at a much slower rate than salmon calcitonin (Otani *et al.*, 1975).

The liver and kidney vascular beds are important sites for the metabolic degradation of calcitonin and there also appear to be heat labile enzymes in the blood which rapidly inactivate the hormone (Munson, 1976; Copp, 1976).

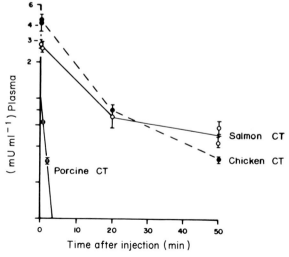

FIG. 19. Disappearance of porcine, salmon and chicken calcitonin after i.v. injection of 5 MRC U kg^{-1} into young rats. (From Copp *et al.* 1972).

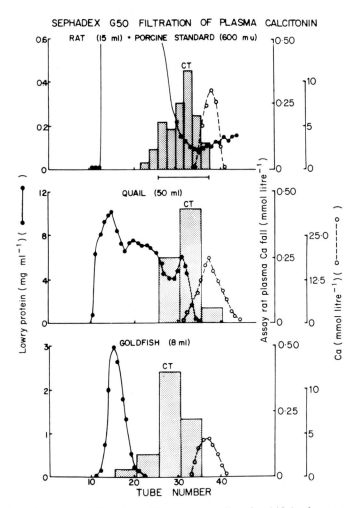

FIG. 20. Sephadex G50 filtration of Japanese quail and goldfish plasma compared with rat plasma to which 600 MRC mU of porcine calcitonin standard had been previously added. Plasma samples mixed with 20% saturated urea solution and filtered on 2·5 × 95 cm columns eluted with 0·1 mol litre^{-1} formic acid, 12 ml fractions. Protein emerged at fraction 13; Ca peaked at fraction 40, urea at fraction 41. Histogram bars represent rat hypocalcaemic bioassay of pooled fractions. Calcitonin (CT) activity confined to fractions 15–35, peak at fractions 31–35 for mammalian and avian samples and at fractions 26–30 for fish sample. Solid line under mammalian data represents fractions normally combined for routine bioassay. (Data from Boelkins and Kenny, 1973 and Dacke, Kenny and Fleming, unpublished observations.)

Studies of plasma calcitonin levels in sub-mammalian vertebrates have been particularly rewarding since in several species, they are sufficiently high to be detected by bioassay. Much of the pioneering work in this area has been carried out by Kenny's group using a combination of a Sephadex method for filtering and concentrating the hormone, followed by hypocalcaemic bioassay in young rats. The relative mobilities of goldfish, Japanese quail and porcine calcitonins in Sephadex G50 columns are shown in Fig. 20. It can be seen that by pooling the fractions containing hypocalcaemic activity and concentrating them by lyophilization, large mol. wt proteins as well as low mol. wt contaminants, which interfere in the bioassay, may be excluded. The maximum levels of plasma calcitonin found by bioassay in several vertebrate classes are represented in Fig. 21. These are particularly high in osteichthyan fish and birds.

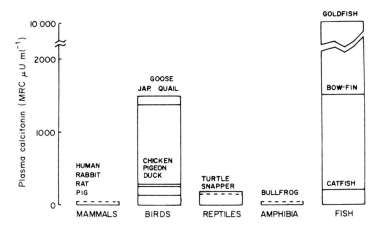

FIG. 21. Vertebrate plasma calcitonin levels; maximum levels found by rat hypocalcaemic bioassay. All mammalian and amphibian levels were undetectable (<50 μU ml^{-1}). In general, birds, reptiles and fish have higher plasma calcitonin levels reaching up to 10 000 μU ml^{-1}. In species such as salmon, even higher levels have been detected by radio-immunoassay. (Mammalian, avian and fish data from Kenny et al. (1972). Additional reptilian and amphibian data Dacke and Kenny, unpublished observations.)

In most other classes, the plasma calcitonin levels are, at best, barely detectable (<50 MRC μ U ml^{-1}), using the techniques outlined above. The extremely high levels of plasma calcitonin found by bioassay in teleosts such as the goldfish (*Carassius auratus*) (Dacke et al., 1971) have been confirmed in the salmon (*Oncorhynchus nerka*) by Copp et al. (1972). In the latter species, maximum levels of 66 MRC mU ml^{-1} were reported.

It is difficult to correlate plasma calcitonin levels in vertebrates with the levels in the gland of origin (see Table X). Although the glandular levels are higher in teleosts and birds than amphibians or reptiles, they are clearly no higher than those in chondrichthyans or mammals. Evidently the high plasma calcitonin levels of fish and birds not only reflect a high specific activity for the homologous hormones, but must also reflect a high rate of secretion.

Further consideration of the relationship between plasma calcitonin levels and calcium regulation in sub-mammals is given in later chapters.

Vitamin D_3 and its Metabolites

Rickets, the deficiency disease of vitamind D_3, has been recognized for several hundred years (De Luca, 1976), but it was only in 1936 that the active steroidal derivative vitamin D_3 (cholecalciferol), was first isolated from fish liver. In the last decade this familiar vitamin has appeared in a new light with the discovery that it is metabolized to active compounds which endocrinologists embrace as hormonal in character. Since the vitamin is either ingested as part of the diet or, in some species, may be synthesized in the skin from a precursor, 7-dehydrocholesterol, in the presence of ultra-violet light, it cannot be said to have a gland of origin. However, the active "hormonal" metabolites are produced in the liver and kidney respectively and as these are familiar organs, their comparative morphology will not be discussed.

The phylogenic distribution of vitamin D_3 amongst vertebrate classes is quite similar to that of calcitonin. The vitamin is present in livers of all extant vertebrate species apart from cyclostomes and chondrichthyans where, at most, trace quantities are found (Urist, 1963). It appears, then, that vitamin D_3 may have evolved at about the same time or slightly later than calcitonin as a Ca regulating factor in vertebrates. As with calcitonin, its appearance seems to correlate with that of bone, perhaps since both substances would seem to have a central function of conserving bone mineral.

CHEMISTRY

Vitamin D_3 and its metabolites are of steroidal origin and are, therefore, lipid soluble. The vitamin has long been delegated a catalytic role in Ca homeostasis; however, its conversion to active forms with subsequent feedback regulation of this metabolism, has only recently been appreciated. This is not surprising since by definition vitamins are catalytic in their action and not, therefore, regulated by feedback. However, in mammals, at least, vitamin D_3 is synthesized from a precursor under the influence of ultra-violet light and not normally ingested as a dietary requirement; it cannot, therefore, be considered a vitamin in the true sense of the word. The chemistry and metabolism of the vitamin as well as its physiological function have been reviewed recently by Omdahl and De Luca (1973) and Norman and Henry (1974). The chemical structures and metabolic relationships of the vitamin and major metabolites are shown in Fig. 22.

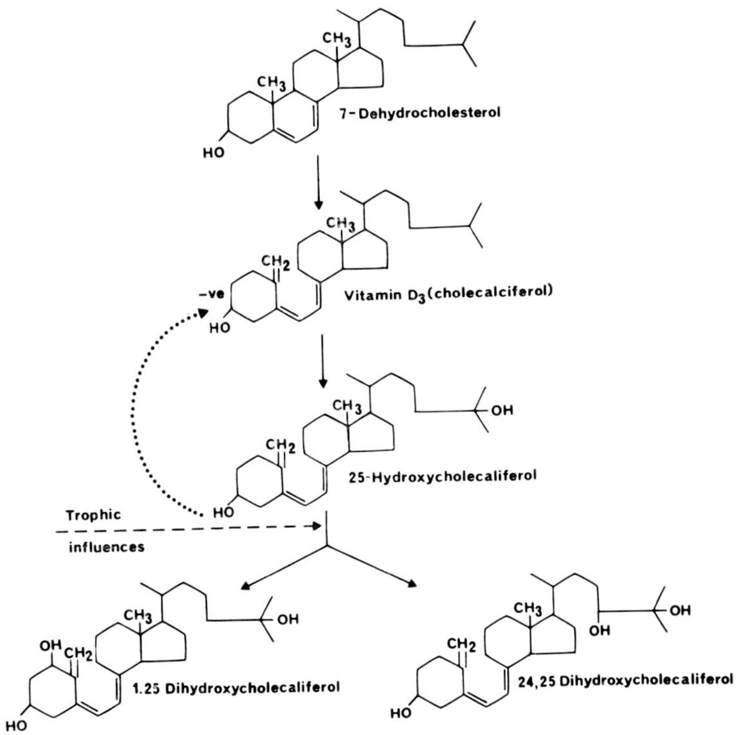

FIG. 22. Vitamin D_3 is formed from the pro-vitamin 7-dehydrocholesterol in the skin of mammals and converted to active metabolites 1,25- and 24,25-dihydroxycholecalciferol. The middle step from cholecalciferol to 25-hydroxycholecalciferol is self-limiting while the final step from 25-hydroxycholecalciferol to dihydroxy derivatives is influenced by a variety of trophic factors (see text).

Thus in mammals and probably in birds, a precursor, 7-dehydrocholesterol is present, either in the skin, or as a secretion of the preen gland respectively. This substance is converted into cholecalciferol which, in mammals, is absorbed directly into the blood stream. In birds, the preen gland oil is spread over the feathers and is then irradiated by ultra-violet light before oral ingestion. This probably forms the basis of familiar behavioural patterns in birds concerned with sunbathing and preening (Sebsell and Harris, 1954). In mammals, cholecalciferol may also be taken in as part of the diet. The next step, a hydroxylation at the 25 position, occurs in the microsomal fractions of liver cells. This step has been demonstrated in rats and chicks (Ponchon et al., 1969). It is self-limiting by a process of negative feedback and it is considered by Omdahl and De Luca (1973) that this bottleneck in vitamin D_3 metabolism helps protect against possible toxic effects which might be caused by the vitamin metabolites and also helps conserve the vitamin when either its dietary intake or formation via ultra-violet irradiation in the skin is low.

A further hydroxylation step occurs in the kidney, probably in the mitochondrial fraction of cells. This normally results in production of 1,25-dihydroxycholecalciferol, a compound considered to be the active form of the vitamin (Frazer and Kodicek, 1970).

The phylogenic distribution of renal hydroxycholecalciferol hydroxylase in vertebrates has been studied by Haussler and Norman and Henry (1974). The presence and activity of this enzyme in different vertebrate classes is shown in Table XII. Unfortunately, these studies did not extend into the two lower vertebrate classes, Agnatha and Chondrichthyes, but hopefully this deficiency in our knowledge will soon be remedied.

When 1,25-dihydroxycholecalciferol formation is depressed, another metabolite 24,25-dihydroxycholecalciferol, may be formed (see Fig. 22). The physiological significance of this is not clear although it is becoming increasingly

TABLE XII

Phylogenic distribution of renal 25-hydroxycholecalciferol-1-hydroxylase

Class	Representative	1,25-dihydroxy cholecalciferol produced (pmol min^{-1} g^{-1} kidney)
Osteichthyes	Goldfish	3
	Rainbow trout	<1
	Carp	78
	Rockfish	5
	Sea bass	7
	Sculpin	2
Amphibia	*Rana pipiens*	8
	Rana macroglossa	13
Reptilia	Iguana	6
	Turtle	56
	Caiman (crococile)	7
	Snake	<1
Aves	Chick (rachitic)	520
	Chick (+D)	140
	Chicken	8
	Duck (rachitic)	390
	Turkey	9
Mammalia	Rabbit	90
	Mouse	<1
	Rat (−D)	90
	Guinea pig (−D)	10
	Cow	2
	Sheep	3
	Bison	16
	Squirrel monkey (−D)	435

(From Norman and Henry, 1974.)

CALCIUM REGULATING HORMONES 65

evident that this metabolite has some biological activity, albeit small, with respect to bone (Atkins, 1976) and gut (Henry et al., 1976).
The fact that the alternative metabolite is produced at all, and that this production appears to be controlled, leads Omdahl and De Luca (1973) to the conclusion that it does have some physiological function.
Further vitamin D_3 metabolites such as 25,26-dihydroxycholecalciferol are also formed, but the function of these, if any, is unknown (De Luca, 1976).

CONTROL OF 1,25-DIHYDROXYCHOLECALCIFEROL PRODUCTION AND SECRETION

We have seen that production of the intermediate metabolite 25-hydroxycholecalciferol in the liver is self-limiting, being controlled by a short loop negative feedback. However, this material is normally removed by further hydroxylation to the active 1,25-dihydroxy or less active 24,25-dihydroxy products. The regulation of this final step is interesting since it is a clear demonstration of the hormonal nature of the end product and we can, therefore, consider the various precursors to be pro-hormones.
At first it was thought the renal hydroxylation step was modulated by the prevailing level of plasma Ca, but this modulation was found to disappear in thyro-parathyroidectomized animals leading to the observation that it is parathyroid hormone which acts as the modulator (Garabedian et al., 1972). Thus parathyroid hormone appears to act in a trophic capacity to stimulate production of 1,25-dihydroxycholecalciferol by the kidney. Experiments involving pre-treatment of chicks in vivo with parathyroid hormone followed by monitoring the conversion of labelled 25-hydroxycholecalciferol to its 1,25-dihydroxy product in vitro are equivocal (Parsons, 1976). Some groups report a positive influence of the hormone on the hydroxylation step while others have reported a negative influence. In the positive experiments parathyroid hormone doses of 30 U.S.P. U per 6 h for two days in one-day-old chicks caused a fourfold increase in 1,25-dihydroxycholecalciferol production. It seems that the concentration of Ca in the incubation medium may be critical in this response. The balance of evidence from several sources, however, is now quite convincingly in favour of parathyroid hormone having a positive trophic influence on 1,25-dihydroxycholecalciferol production by the kidney under physiological conditions. There is evidence that calcitonin can inhibit formation of the metabolite (Rasmussen et al., 1972), although this is not considered to be as important an influence as that of parathyroid hormone (Omdahl and De Luca, 1973). There is also evidence for low cellular inorganic phosphate concentrations stimulating 1,25-dihydroxycholecalciferol production (Omdahl and De Luca, 1973), a factor to be taken into account when considering the evolutionary significance of vitamin D_3 and its role in bone metabolism (see Chapter 14).
Another hormone which appears to be trophic for 1,25-dihydroxycholecalciferol production is prolactin. In pregnant and lactating mammals, Ca and phosphate absorption are greatly increased (Horrobin, 1974), but until re-

cently the physiological mechanism of this response remained obscure. It was shown that following prolactin treatment, production of the vitamin D_3 metabolite by chick kidney homogeonates is increased. A single subcutaneous injection of 70 μg ovine prolactin into one-day-old chicks resulted in a doubling of the renal 25-hydroxycholecalciferol-1-hydroxylase activity within 1 h. The enzyme activity could be boosted in both vitamin D_3 deficient and D_3 supplemented chicks (Spanos et al., 1976). This group has also developed a method for measuring circulating 1,25-dihydroxycholecalciferol using a radioreceptor assay. When given injections of 100 μg ovine prolactin per day for 5 days, plasma levels of the metabolite were increased from 15·9 ng 100 ml^{-1} –28·6 ng 100 ml^{-1}, a finding consistent with the previous observation on renal 25-hydroxycholecalciferol-1-hydroxylase activity (Spanos et al., 1976). These authors have suggested that the results in chicks might be extrapolated to mammals in order to explain the increase in Ca and phosphate absorption during late pregnancy and lactation. Gonadal hormones also appear to act in a trophic capacity for 25-hydroxycholecalciferol hydroxylation in a wide range of vertebrates. Thus in bullfrogs (*Rana catasbeinae*) the normal rate of 1,25-dihydroxycholecalciferol production is 0·49 pmol min^{-1} g^{-1} kidney, while the corresponding value for 24,25-dihydroxycholecalciferol is 0·73 pmol min^{-1} g^{-1}. Intramuscular injection of oestradial benzoate (3 mg kg^{-1}) for 6 days changed these rates to 0·93 and 0·13 pmol min^{-1} g respectively (Kenny et al., 1977). Similar data have been obtained with respect to oestrogen and vitamin D_3 metabolism in birds (Kenny, 1976) and mammals (Baksi and Kenny, 1977). The avian data are discussed in connection with egg-lay in birds in Chapter 13. It is not yet clear whether the influence of oestrogens are mediated via prolactin secretion, since in mammals the latter hormone is released by an oestrogen stimulus (Nicoll et al., 1962). It seems, therefore, that 1,25-dihydroxycholecalciferol and possibly 24,25-dihydroxycholecalciferol can be considered as hormonal in character with respect to their synthesis and release, since these functions are clearly regulated by a wide variety of trophic factors. On this basis we must conclude that the precursor, 25-hydroxycholecalciferol and even the more primary metabolites are pro-hormonal in character. This is a situation which appears to be unique amongst the vitamins.

CIRCULATING LEVELS OF VITAMIN D_3 METABOLITES

Cholecalciferol and 25-hydroxycholecalciferol are transported in the blood in association with plasma proteins (Hay and Watson, 1977). These range from albumens in birds and mammals to α globulins in osteichthyans, reptiles, some birds and mammals and β globulins in other birds. For amphibians, agnathans and chondrichthyans, the carrier molecules are apparently lipoproteins, although whether they have this function in the two latter classes is a matter for conjecture. Presumably similar transport mechanisms exist for 1,25-dihydroxycholecalciferol and 24,25-dihydroxycholecalciferol in the vertebrate circulation, but these have yet to be elaborated.

Assays for circulating 1,25-dihydroxycholeciferol in human and chick plasmas are available (Brumbaugh et al., 1974a, b; Spanos et al., 1976) and, hopefully, will soon be applied to other vertebrate classes. In human plasmas these assays give concentrations for the metabolite of 4 ng 100 ml^{-1} while the circulating level in chick plasma is around 8 ng 100 ml^{-1}.

The Corpuscles of Stannius and Hypocalcin

A fourth Ca regulating system in vertebrates which is gaining attention comprises the corpuscles of Stannius.

Morphology

These glandular structures are located posteriorly on the kidney or its ducts. They are only found in holostean and teleost fish, and then not in all species. In the Salmonidae in particular, the corpuscles are absent. Their number varies from 1 to 3 pairs in teleosts while the holostean, *Amia calva* has 20 to 25 pairs. Chondrostean fish such as the sturgeon (*Acipenser fulvescens*) lack the corpuscles (Krishnamurthy and Bern, 1969). These authors have suggested that the large numbers of corpuscles found in holosteans and their apparent decline in teleosts may reflect an evolutionary trend in that they have less functional importance in the latter class.

Embryologically the corpuscles of Stannius may be derived from the pronephric duct or, more rarely, the mesonephric duct (Bentley, 1976).

Nature of Active Substances from Corpuscles of Stannius

The cytological architecture of the corpuscles marks them as protein-secretory, rather than as steroid producers (Maetz, 1968). He has suggested that they subserve three functions independent of pituitary control: Involvement with interrenal corticoids to regulate electrolyte balance; a source of renin to control blood pressure and kidney function; and in the control of calcaemia. This latter function is of interest.

In eels (*A. anguilla*) adapted to both hypercalcic and hypocalcic media, Stanniectomy results in transiently increased plasma Ca levels and hypophosphatemia (Fontaine, 1964; Chan et al., 1967). Several other groups of workers have repeated these results for various species of fish (Simmons, 1971). Normal blood Ca levels may be restored by injection of extracts from the extirpated organs or by reimplantation of the glands.

Preliminary attempts at extracting and assaying the active hypocalcaemic principle from fish corpuscles of Stannius have been made by Pang et al. (1974) who tested saline homogeonates of the corpuscles from cod (*Gadus morphua*) or killifish (*Fundulus heteroclitus*) for hypocalcaemic activity when reinjected into killifish. Killifish were most sensitive to these injections if maintained in Ca deficient media for six weeks prior to the assay (see Fig. 23). Even so, four daily injections of the homogeonate were needed to produce a response. With killifish extract, the amount of material needed per injection to produce a threshold response was equivalent to one corpuscle, while ten times this quantity was needed for a maximal response. It is obvious that large quantities of corpuscles would be needed for just one assay. It would seem premature to consider using the method at its present level of sensitivity for

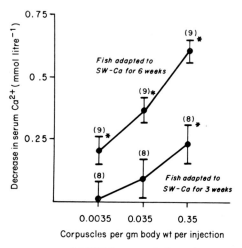

FIG. 23. Log-dose response of killifish to cod Stannius corpuscle homogenate. (*) sig. p <0·05. (From Pang et al., 1974.)

studies on step by step purification of the corpuscles, although hopefully this will be accomplished eventually. In their studies, Pang et al. (1974) found that boiling the homogeonates for 30 min rendered them inactive, thus providing more evidence for the proteinaceous nature of the active material. Furthermore, salmon calcitonin, cortisol or angiotensin had no hypocalcaemic effect in the assay; indeed, the latter substance produced significant hypercalcaemic responses. It is clear from these results that the active hypocalcaemic principle in corpuscles of Stannius is distinct from calcitonin or any sodium or blood pressure regulatory substances. Pang and his coworkers have, therefore, proposed the name "hypocalcin" for the hypocalcaemic factor.

REGULATION OF STANNIUS CORPUSCULAR HYPOCALCAEMIC ACTIVITY

The activity (from histological observations) of the corpuscles of Stannius in killifish appears to be greater when the fish are maintained in normal sea water than if they are in artificial sea water with a low Ca concentration (Pang et al.,

1973). In normal sea water the corpuscles exhibit rough endoplasmic reticulum, hyperactive Golgi apparatus, granular depletion and the presence of lysosome-like bodies, all indicative of synthetic and secretory activity. These indications of activity are not evident in corpuscles from fish maintained in Ca deficient sea water.

The Pituitary Gland

In vertebrates there is evidence that the pituitary plays an important role in Ca regulation. Its function in fish is apparently similar to that of the parathyroids in tetrapods. When killifish (*F. heteroclitus*) maintained in Ca free artificial sea water are hypophysectomized, they become liable to tetanic contractions of the muscles similar to those of higher vertebrates following parathyroidectomy. At the same time, a considerable decline in plasma Ca levels occurs. If the fish are maintained in high Ca sea water they do not exhibit these responses. Furthermore, the response in low Ca, hypophysectomized fish can be reversed by replacement therapy with fish pituitary extracts (Pang, 1973).

Whether these responses reflect the activity of well-known pituitary hormones such as prolactin and corticotrophin which play a major role in fish hydromineral balance (Bentley, 1976), or whether separate hypercalcaemic pituitary factors exist, remains to be seen. However, hypercalcaemic effects of corticotrophin, prolactin and cortisol (which could be under pituitary control) have all been demonstrated in fish (see Table XIII). Thus Pang *et al.* (1973) consider prolactin to be the most effective of these fish hypercalcaemic substances. Indeed, prolactin appears to have hypercalcaemic effects in a wide variety of vertebrate classes including teleost fish, birds, possibly amphibia (Nicoll and Bern, 1972) and mammals (Horrobin, 1974). Perhaps these hypercalcaemic effects reflect a trophic action of prolactin on renal 25-hydroxycholecalciferol-1-hydroxylase activity similar to that demonstrated in birds by Spanos *et al.* (1976a, b).

TABLE XIII
Effects of pituitary hormones on serum Ca levels of hypophysectomized male *F. heteroclitus* adapted to Ca deficient sea water and fed low Ca food

Groups	Ca (mmol litre^{-1}) $\pm$ s.e.
(1) Operated controls + saline	2·36 $\pm$ 0·15 (5)[a]
(2) Hypophysectomized + saline	1·75 $\pm$ 0·13 (5)
(3) Hypophysectomized + cortisol	2·18 $\pm$ 0·06 (10)[a]
(4) Hypophysectomized + ACTH	2·50 $\pm$ 0·27 (10)[a]
(5) Hypophysectomized + prolactin	2·49 $\pm$ 0·14 (5)[a]
(6) Hypophysectomized + MSH	1·97 $\pm$ 0·08 (10)

[a] Significantly different from group 2 ($p < 0.05$).

Prolactin type molecules are found in the most primitive vertebrates, unlike some of the other pituitary hormones which are a later development. Indeed there is some evidence that prolactin may be the oldest of pituitary hormones (Horrobin, 1973), although Bentley (1976) suggests that its presence in cyclostomes is equivocal. Prolactin has some functional and chemical homology with growth hormone which is also thought to be primitive (Bentley, 1976).

Recently Parsons et al. (1978) demonstrated a separate hypercalcaemic principle in fish pituitary glands which appears to have a close structural homology with mammalian parathyroid hormone. The substance extracted from cod (*G. morhua*) and eel (*A. anguilla*) pituitary glands was immunologically similar to bovine and human 1-34 parathyroid hormone, although the displacement curves for these extracts were not parallel with those of bovine 1-84 standard. The fish pituitary extracts had significant hypercalcaemic activity within two hours of injection into intact trout (*Salmo gairdneri*), a much faster response time than that found with prolactin. Furthermore, the hypercalcaemic activity coeluted with mammalian parathyroid hormone following gel filtration. Mammalian parathyroid hormone was without activity, however, in the trout system. Parsons and his coworkers have also extracted similar activity from pituitaries of vertebrates as high as the rat, although not from human glands, and consider the *pars intermedia* to be the source of this activity since humans do not have this pituitary structure. They also speculate that the material may be present in the primitive cyclostomes. In order to distinguish this material from the hormone of the parathyroid glands, Parsons and his coworkers have suggested the name "hypercalcin" for the pituitary factor.

Gonadal Hormones

In the sub-mammals, the gonads have influences on Ca metabolism which are often more obvious than the mainly skeletal effects to be found in mammals. Oestrogens facilitate the formation of a phospholipoprotein (vitellin) in the liver of reproductively active females ranging from the class Osteichthyes to the Aves (Simkiss, 1967; Bentley, 1976). Vitellin circulates in the blood and is subsequently incorporated into the yolk of the developing egg. This plasma protein binds Ca avidly and its presence is usually associated with an increased plasma Ca concentration (see Table XIV). The reaction is not generally found in mammals or in the lower classes, Agnatha and Chondrichthyes. It is probable that this response reflects the needs of vertebrates producing large megalecithal eggs, and aids in the Ca requirement of the developing embryo.

Both oestrogens and androgens play a role in Ca metabolism of egg-laying birds in which they are involved in formation and maintenance of medullary bone (see Chapter 4). Also mentioned previously in this chapter is the evidence for a relationship between oestrogens and vitamin D_3 metabolism which could affect bone and gut Ca metabolism.

TABLE XIV
Blood Ca levels and hormonal status in vertebrates

Class	Species	Skeleton	Blood Ca (mmol litre⁻¹) Normal male	Blood Ca Oestrogen treated	Vitamin D	Hormonal status[a] PTH	CT	PRL	HC	Vitamin D	Hormonal effects[b] PTH	CT	PRL	HC
Agnatha	Hagfish	Cartilage	5.1	5.3	±	−	−	±	−	−	−	−	−	
Chondrichthyes	Shark	Calcified cartilage	4.3	4.1	±	−	+	+	−	−	−	−	−	+
Osteichthyes	Bass	Bone	3.1	27.0[c]	+	−	+	+	+	+	+	+	+	
Amphibia	Frog	Bone	2.1	9.2[c]	+	+	+	+	−	+	+	±	+	
Reptilia	Turtle	Bone	2.9	39.0[c]	+	+	+	+	−	+	±	±	+	
Aves	Chicken	Bone	2.5	23.0[c]	+	+	+	+	−	+	+	+	+	
Mammalia	Mouse	Bone	2.4	2.8	+	+	+	+	−					

Parathyroid hormone (PTH), calcitonin (CT), prolactin (PRL), hypocalcin (HC).
[a] Indicating presence (+) or absence (−) of hormone or gland.
[b] Indicating report of a positive (+) or negative (−) response of blood Ca to either injection of hormone or removal of its source.
[c] Significantly different ($p < 0.05$) from normal male levels.
(Modified from Kenny and Dacke, 1975.)

Other Hormones and Controlling Influences

The hormones which have been considered so far are the major ones thought to play a role in vertebrate Ca metabolism. Many other hormones can influence growth and turnover of the skeleton, often by virtue of their general anabolic or catabolic effects. These include such hormones as pituitary somatotrophin (growth hormone), the thyroidal hormones, interrenal hormones (including previously mentioned cortisol) and those of the pineal (Vaughan, 1970). It is unlikely that any of these play more than a minor role in short-term Ca regulation under normal physiological conditions.

Another factor often mentioned in studies of bone metabolism is vitamin A. Large doses of the vitamin increase bone resorption *in vitro*, an effect which may be counteracted by calcitonin (Vaughan, 1970). There is no evidence that the vitamin exerts a controlling influence in the same way as vitamin D_3 or that it undergoes a complex metabolism to active hormonal metabolites as does vitamin D_3. However, it is a fact that large quantities of vitamin A, like D, are stored in the liver of fish (Urist, 1976b) and in species such as chicks, derangement of bone formation and growth occur if dietary sources of the vitamin are deficient (Coates, 1971).

6. Target Organs in Calcium Homeostasis

General

In the previous chapter an account is given of the nature of endocrine systems involved in vertebrate calcium (Ca) regulation. In this chapter, responses to these hormones of a variety of target organs will be considered. Target organs often respond to several different hormones which can have antagonistic or synergistic actions at these sites and this is certainly the case with the regulation of the Ca ion. Current thinking in the role of hormones and their target tissues in mammalian Ca regulation is summarized in Fig. 24.

In sub-mammals, there are, as we have seen, several variations in the type and nature of Ca regulating hormones. Similarly there are differences in the

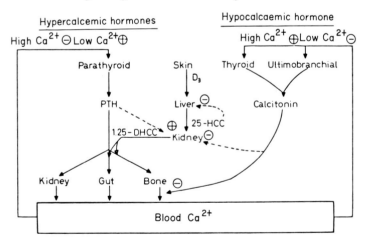

FIG. 24. A current concept of the endocrine control of Ca homeostasis in mammals. Inhibitory effects ($\ominus$), stimulatory effects ($\oplus$). Parathyroid hormone (PTH), vitamin D_3 (D_3), 25-hydroxycholecalciferol (25-HCC), 1,25-dihydroxycholecalciferol (1,25-DHCC). The kidney is emerging as the integrating organ responsible for conversion of 25-HCC to its active metabolites 1,25-DHCC. This reaction appears to be under control of positive trophic influences such as PTH and prolactin, with calcitonin possibly having an inhibitory influence. (From Kenny and Dacke, 1975.)

target organs, which have to be identified. In mammals, the three major target tissues with respect to Ca regulation are considered to be bone, gut and kidney. In the sub-mammals, additional target organs may include gills, skin, endolymphatic sacs and scales and, in special circumstances, the oviduct.

Bone

The processes of bone accretion and resorption are known collectively as "bone turnover" and normally, in the adult animal, these processes balance each other out. In young, growing animals, accretion usually predominates, while in certain adult conditions such as pregnancy or egg-lay, net resorption may occur for limited periods of time. Bone is influenced by a wide variety of Ca regulating hormones.

PARATHYROIDS AND BONE

Until fairly recently it was considered that parathyroid hormone had a mainly catabolic (resorptive) effect on bone which contributes towards its hypercalcaemic action. It is now becoming clear that the hormone also has an important anabolic effect on bone which, under normal physiological conditions, Parsons (1976) considers to be even more important than its catabolic effect.

Parathyroid hormone has a wide variety of actions on bone cells including increasing osteoclastic activity, osteocytic osteolysis and changes in osteoblastic activity (see Vaughan, 1970); both rapid and slow effects on bone cells have been demonstrated (Kenny and Dacke, 1975; Gray et al., 1974). Cellular Ca pumps are probably involved in the fast responses to parathyroid hormone in which Ca is moved from the bone fluid compartment to the extracellular fluid and Gray et al. (1974) consider that these Ca movements constitute the important mechanism by which parathyroid hormone can rapidly provide Ca for the blood. It seems likely that these pumps are located in the osteocytes.

According to Parsons et al. (1971) the most immediate effect of parathyroid hormone is to promote movement of Ca ions into bone cells, thus causing a transient hypocalcaemia. This response occurs in mammals and also in egg-laying chickens provided they are fed high (5%) Ca diets. If immature chickens or Japanese quail are fed more normal (2·3%) Ca diets they do not show the initial hypocalcaemia (Mueller et al., 1973; Kenny and Dacke, 1974). It is suggested that these initial, and somewhat elusive, hypocalcaemic responses may reflect movement into bone and soft tissue cells of Ca ions which then act as a second messenger to activate 3′,5′-cyclic AMP mediated Ca pumps (Kenny and Dacke, 1975). These then pump Ca ions away from the resorbing bone surface, probably by activating a phosphorylase kinase which, in turn, activates more components of one or more enzyme cascades (Parsons, 1976; Rasmussen et al., 1976).

Other actions of parathyroid hormone include an increase in production of organic acids within bone. Lactic, citric and carbonic acids have all been implicated in the solubilization of bone mineral by decreasing local pH, these acids probably being produced by osteoclasts. Presumably one way in which a fall in pH could mobilize bone would be by acidification of:

$$PO_4 \rightarrow HPO_4 \rightarrow H_2PO_4$$

which, in turn, would cause a fall in the Ca × P solubility product. Parathyroid hormone produces general increases in cell metabolic activity and an increase in at least the osteoclast population on resorbing bone surfaces (Kenny and Dacke, 1975).

The responses outlined above can be produced by injections of large (pharmacological) doses of parathyroid hormone. In normal physiology it is unlikely that such high levels of circulating hormone would be found except under acute or pathological situations. The question is, whether or not experimental responses to high doses of hormones still occur with the lower levels that are found under normal physiological conditions.

Parsons (1976) in reviewing the available evidence, considers that under physiological conditions the most important bone responses to parathyroid hormone are anabolic rather than catabolic. These responses to chronic and often low dosage of parathyroid hormone included increases in osteoblastic activity, alkaline phosphatase and incorporation of various labels into bone and, most convincingly, increases in the mass and mineral content of bone. It is of interest that a majority of papers cited by Parsons (1976) involved experiments with young animals in which the skeleton would still be growing. Such anabolic responses are not inconsistent with the traditional hypercalcaemic role of parathyroid hormone, but lend support to those workers who consider soft tissues (gut and kidney) to be major mediators of the hypercalcaemic responses. In addition to its mineral regulating role, then, we should consider parathyroid hormone as a regulator of skeletal metabolism.

Parathyroid hormone probably has similar influences on bone in all vertebrates which possess this tissue; even fish, which apparently lack the hormone, are able to respond to injections of mammalian parathyroid extract. Thus Budde (1958) who treated guppies (*Lebistes reticulatus*) with parathyroid extract, reported osteoblastic modulation to fibroblasts and some signs of enhanced osteogenesis, these effects bearing a marked similarity to those of small doses of the hormone in rats. Other workers (Fleming and Meier, 1961; Clark and Fleming, 1963; Fleming, 1967) have demonstrated hypercalcaemia and hyperphosphaturia in killifish (*Fundulus kansae*) injected with parathyroid extract; again, these responses are similar to those of mammals. In marine sharks and rays, however, intramuscular doses of 100 U.S.P. U of hormone had no effect on plasma Ca levels (Urist, 1962). These findings give credence to the suggestion discussed in Chapter 5, that parathyroid hormone may have evolved from a pituitary hormone in pre-amphibious vertebrates, a theory supported by the structural homology which parathyroid hormone shares with corticotrophin (Parsons, 1976). Presumably such an archaic hormone would have had receptor sites in bone and other target tissues associated with Ca regulation.

CALCITONIN AND BONE

Calcitonin produces rapid hypocalcaemic and hypophosphataemic effects in enterectomized, nephrectomized, young mammals. However, it is inactive in rat embryos before the onset of ossification (Simmons, 1976). Evidence from studies *in vivo* and *in vitro* have long established bone as the main target organ for this hormone (Gray et al., 1974). Calcitonin appears to inhibit both osteoclastic resorption and osteocstic osteolysis, while there is some evidence that it may also stimulate osteoblaytic accretion (Simmons, 1976). Thus chick embryonic bones *in vivo* show evidence of enhanced bone weight and mass and also increased bone alkaline phosphatase following injections of up to 750 MRC mU of porcine calcitonin (McWhinnie, 1975). The hormone also blunts typical parathyroid hormone mediated responses such as increased production of lactic and citric acids, pyrophosphatase and lyosomal enzyme production, serum alkaline phosphatase and urinary hydroxyproline; these all indicate an action of calcitonin which is proportional to the prevailing rate of bone remodelling (Simmons, 1976).

The inhibition of resorption by calcitonin in cultured mouse foetal bones previously stimulated with parathyroid hormone or vitamin D_3 metabolites, is rather transient (Brand and Raisz, 1972), while continuous bombardment with high levels of calcitonin results in an apparent loss of response to this hormone (Wener et al., 1972). This phenomenon of "escape" is encountered frequently in endocrine systems. It appears that escape from the influence of calcitonin only occurs in bone in which resorption is being strongly stimulated. This has implications in sub-mammalian species, particularly osteichthyan fish, and birds, which have relatively high levels of circulating calcitonin (see Chapter 5). Are the bones of these species under permanent escape from calcitonin, with the hormone only exerting its effects on other target organs, or are the bones protected by calcitonin from resorptive agents such as parathyroid hormone or vitamin D_3? In any case, since parathyroid hormone only appears to have transitory effects in those sub-mammalian classes in which it is present, sustained bone resorption is not as likely to occur in these classes as in mammals.

The action of calcitonin on bone cells appears to involve activation of adenyl cyclase (Heersche et al., 1974) and thus $3',5'$-cyclic AMP production, while Ca efflux from these cells is inhibited according to Harell (1973). It is considered by Talmage and his colleagues that the inhibition of Ca efflux from bone cells is secondary to an increased influx of inorganic phosphate ion as indicated by the behaviour of acutely and chronically injected ^{32}P label (Kennedy and Talmage; 1971, Talmage et al., 1972). Other experiments conflict with those cited above as they indicate that calcitonin may actually increase Ca efflux from bone cells, thus lowering the concentration of active Ca ion within them (Parsons, 1976). This sequence of events would probably have the opposite effect from that of parathyroid hormone on the adenyl cyclase system, since that hormone initially raises cellular Ca which then acts in a secondary messenger capacity (Parsons, 1976).

It is clear that further experiments are needed in order to establish the subcellular mechanisms by which parathyroid hormone and calcitonin influence bone. It is not yet apparent if the adenyl cyclases activated by these hormones are identical or lie at different sites within bone cells (Rasmussen et al., 1976) although there is an indication from studies of 3',5'-cyclic AMP production in isolated calvariae, that these sites could be different. These studies suggest that the effects of parathyroid hormone and calcitonin are additive even at high concentrations, a situation which might be expected considering the antagonistic effect which these hormones have with regard to bone Ca turnover (Aurbach and Chase, 1976). Furthermore, it has been demonstrated that the parathyroid hormone sensitive adenyl cyclase system in rat foetal calvariae can be inactivated by trypsin proteolysis, while the calcitonin sensitive adenyl cyclase in the same preparation is not affected (Chase and Obert, 1975). These authors suggest that the selective inhibition of the parathyroid hormone sensitive system is evidence for distinct receptor sites for the two hormones.

In addition to its effect on bone resorption, there is evidence that calcitonin may increase the rate of bone formation in that skeletons of animals undergoing chronic dosage with the hormone increase their bone mass, Ca content, ^{45}Ca retention, tetracycline labelling, osteoblastic activity, alkaline phosphatase and pyrophosphatase activities (Simmons, 1976). These responses may, in part, reflect a rebound from the effect of parathyroid hormone on the bone, particularly as they are generally found after chronic rather than acute calcitonin treatment.

An important feature of the acute action of calcitonin in mammals is its age dependency, the duration and amplitude of the hypocalcaemic and hypophosphataemic responses becoming less pronounced in older animals. This reflects a reduction in the prevailing rate of bone turnover in the more mature animal. For the same reason, males, which have a longer growing period than females, are slightly more responsive to the hormone (Simmons, 1976).

Vitamin D_3 Metabolites and Bone

The fact that vitamin D_3 brings about normal calcification in bone has led many investigators to consider a direct function of the vitamin on bone accretion, but it is clear from various studies (Omdahl and De Luca, 1973) that this does not occur. The major effect of the vitamin, or at least its metabolite 1,25-dihydroxycholecalciferol, on bone is, paradoxically, to bring about its resorption and hence contribute towards the hypercalcaemia produced by this substance.

At present, little is known of the biochemical mechanism by which 1,25-

dihydroxycholecalciferol induces bone resorption, although it is clear that resorption does occur (Wong et al., 1972); thus labelled 1,25-dihydroxycholecalciferol is concentrated into bone cell nuclei of rachitic chicks (Weber et al., 1971). It is also clear (Norman and Henry, 1974) that inhibitors of DNA directed RNA synthesis, such as actinomycin D, block vitamin D_3 induced bone resorption. These findings indicate that 1,25-dihydroxycholecalciferol may interact with bone cells in a similar fashion to that with gut mucosal cells (see following section), resulting in increased production of a Ca binding protein which mediates Ca efflux from bone fluid to extracellular fluid. This concept needs substantiation, however, by direct experimentation.

It has long been conjectured that parathyroid induced bone resorption is vitamin D_3 dependent; the two substances are now seen to have similar but independent effects in bone (Raisz, 1976). Thus parathyroid hormone can resorb bone in vitamin D_3 deficient animals, provided that plasma Ca and inorganic phosphate levels are maintained (Omdahl and De Luca, 1973) and calcitonin can inhibit bone resorption in the presence or absence of the vitamin (Morii and De Luca, 1967).

The alternative vitamin D_3 metabolite, 24,25-dihydroxycholecalciferol, does not appear to be active with respect to bone resorption in either chicks or rats according to Henry et al. (1976). However, Atkins (1976) finds it a potent stimulator of bone resorption in tissue culture and also that it enhances the effect of the 1,25-dihydroxy metabolite in this system.

OESTROGEN AND BONE

The most obvious effects of oestrogens on bone occur in birds in which large quantities of intramedullary bone are formed during the pre-ovulatory phase of the egg-laying cycle. Medullary bone can also be artificially induced in cockerels by injections of oestrogen, or in young birds by combined injection of oestrogen and androgen (see Chapter 4). Very high chronic dosage with oestrogens may also affect rodent bones. These strictly pharmacological responses involve a suppression of osteoclastic resorption, and in mice, at least, an increase in endosteal osteoblast production, resulting in obliteration of marrow spaces by spicules of unresorbed trabecular bone (Simmons, 1976).

Most of the evidence for an oestrogen effect, at least in mammalian bone, is indirect. Women develop osteoporosis following the menopause, a disease which can be arrested but not reversed by treatment with oestrogens and androgens (Raisz, 1976). Possibly some of the actions of gonadal hormones in bone are mediated indirectly by their effect on plasma calcitonin levels or via vitamin D_3 metabolism (see Chapter 5).

THE PITUITARY AND BONE

While it seems clear that hormones such as growth hormone, thyroid hormones etc. exerting general anabolic or catabolic effects, will have an influence on skeletal mass and development, little is known about possible effects of those pituitary hormones with specific calcaemic actions, on the bone. It seems clear that prolactin has important hypercalcaemic influences which, in fish, may be mediated in part by the gills (see below and Chapter 5). This hormone also plays a trophic role in the production of vitamin D_3 metabolites (see Chapter 5), and is, therefore, likely to influence bone indirectly via this system.

Recent data from Brehe and Fleming (1976) indicate a pituitary influence on Ca distribution between the scales and bone in teleost fish; furthermore, their data suggest a role for more than one pituitary hormone in these tissues (see Chapter 10).

Gut

The gastrointestinal tract plays an important role in vertebrate Ca regulation and in the tetrapods, from reptiles upwards, ingested food and drinking water are the sole sources of environmental Ca. Fish and amphibia are able to absorb part of their Ca requirement across the skin and gill epithelia. Since Ca can be both absorbed from; and secreted into the gut, the organ makes an ideal target for the action of Ca regulating hormones.

Gut mucosal microvilli increase the surface area for absorption while there is also evidence for loose junctions at the apical region of adjacent cells when compared with the relatively tight junctions of other absorptive tissues such as frog skin. General mechanisms of Ca and phosphate transport have recently been reviewed by Wasserman and Taylor (1976); these include active transport systems capable of pumping Ca against an electrochemical gradient from lumenal to mucosal surfaces, as well as passive diffusion down chemical gradients. The same authors discuss recent evidence for separate phosphate active transport mechanisms but these have not been so widely studied as the Ca transporting system.

PARATHYROID HORMONE AND GUT

Many workers now consider that parathyroid hormone has no direct action on the enhancement of Ca uptake by the gut, although it has a major indirect influence via regulation of vitamin D_3 metabolism which, in turn, affects gut (Kenny and Dacke, 1975). However, evidence for a direct and specific effect of the hormone on intestinal Ca absorption has been provided by Olson *et al.* (1972a), using an *in vitro* vascularly perfused rat intestinal preparation. Enhanced ^{45}Ca transport from mucosal to serosal surfaces was observed within 30 min of infusing 100 U.S.P. U of highly purified parathyroid hormone; while this was apparently a direct response, the dose of hormone was very high and it is doubtful that the effect occurs in normal physiology.

Parathyroid hormone may also influence magnesium transport by the intestine. Clark and Rivera-Cordero (1973) stimulated endogenous parathyroid hormone secretion in rats by feeding low Ca diets which resulted in an increased Ca and decreased magnesium absorption, while phosphate absorption was unaffected. Parathyroidectomized rats did not exhibit these responses. Whether or not the inhibition of magnesium absorption is direct or mediated by increased renal 1,25-dihydroxycholecalciferol production, remains to be seen.

CALCITONIN AND GUT

Using the *in vitro*, vascularly perfused rat intestinal preparation, Olson et al. (1972a, b) have studied the influence of calcitonin on this tissue. Relatively small doses of the hormone (10 MRC mU ml^{-1}) inhibited Ca transport in vitamin D_3 replete intestine, but not in vitamin D_3 deficient ones; higher doses which might be considered supraphysiological in mammals (500 MRC mU ml^{-1}) stimulated vitamin D_3 induced transport. The explanation of these results is not clear, but apparently, the effects are not mediated via inhibition of renal 1,25-dihydroxycholecalciferol production. Presumably the lower dose of calcitonin inhibits the 1,25-dihydroxycholecalciferol induced gut Ca transport system, an effect analogous to the inhibitory effects of calcitonin in bone.

Preliminary evidence in frogs suggests that calcitonin may also modulate Ca transport in the gut of sub-mammalian classes. This evidence is reviewed in Chapter 11. There is little evidence in support of a role for calcitonin in inorganic phosphate transport by the intestine (Wasserman and Taylor, 1976).

VITAMIN D_3 METABOLITES AND THE GUT

Gut is a major target organ for the action of vitamin D_3 metabolites in mammals, birds and, probably, other vertebrates. In these animals, 1,25-dihydroxycholecalciferol plays a central role in the translocation of Ca from the lumenal to serosal surface of gut mucosal cells. It is now clear that the very slow responses (30–50 h) in Ca uptake following injection of native cholecalciferol can be accounted for by the complex metabolism of this substance to its active form (see Chapter 5). If a dose of 1,25-dihydroxycholecalciferol is injected into a rat or chick, the *in vivo* increase in gut Ca absorption reaches a peak 12–15 h later. The mechanisms by which the metabolite increases intestinal Ca uptake have been reviewed by Norman and Henry (1974) and Wasserman and Taylor (1976). The main events in 1,25-dihydroxycholecalciferol mediated stimulation of Ca translocation by gut mucosal cells are summarized in Fig. 25.

These events include an increase in protein synthesis with resulting production of a specific binding protein for Ca (CaBP) having a mol. wt 8200. Increases in ATPase and alkaline phosphatase also occur; these processes have

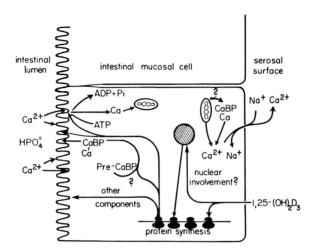

FIG. 25. Schematic summary of possible function(s) of 1,25-dihydroxycholecalciferol (1,25-$(OH)_2D_3$) in Ca translocation in the intestinal mucosa cell. (After Omdahl and De Luca, 1973.)

been demonstrated in both mammalian and avian intestine (Norman and Henry, 1974), and presumably also occur in the intestines of lower vertebrate classes.

It is not clear at present how Ca binding protein is involved in Ca transport, but this process requires ATP (Wasserman and Taylor, 1976).

Vitamin D_3 or its metabolites also enhance absorption of inorganic phosphate in rat and chick gut; this seems to be a primary response to the vitamin system rather than a secondary response related to increased Ca absorption (Wasserman and Taylor, 1976). While the alternative vitamin D_3 metabolite 24,25-dihydroxycholecalciferol has similar effects on gut as 1,25-dihydroxycholecalciferol, it only has about 10–20% of the activity of the latter metabolite, at least in chick or rat intestinal preparations (Henry et al., 1976).

Kidney

The kidney has a dual role in vertebrate Ca regulation. As we have seen, it is involved in the synthesis and secretion of the active vitamin D_3 metabolite and thus functions as an endocrine gland. It also acts as a target organ for Ca regulating hormones and some workers consider kidney to be the major target organ, at least with respect to mammalian Ca regulation.

All vertebrates possess functional kidneys, although these vary greatly in structure from class to class. In fish, part of the function of kidneys is taken over by the gills but kidneys assume greater importance as organs of excretion in tetrapods.

Parathyroid Hormone and Kidney

The mammalian kidney reacts promptly to i.v. injection of parathyroid hormone by increasing the urinary excretion of phosphate, sodium, potassium and bicarbonate ions and decreasing excretion of hydrogen, ammonium, Ca and magnesium ions (Vaughan, 1970). The most notable of these effects is on phosphate excretion which results from an inhibition of tubular phosphate reabsorption. The response is very rapid but short-lived and is preceded by an increase in urinary 3′,5′-cyclic AMP level (Fig. 26) according to Chase and Aurbach (1968).

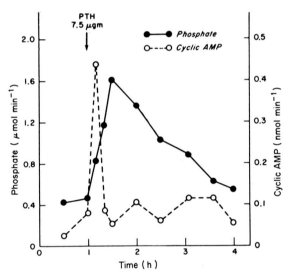

FIG. 26. The effect of parathyroid hormone on the excretion of phosphate and 3′,5′-cyclic AMP by a parathyroidectomized rat. PTH was injected i.v. over a 2 min period. Note the enhanced excretion of both 3′,5′-cyclic AMP and phosphate and, more importantly, the rise in excretion of the former preceding the excretion of the latter. (From Chase and Aurbach, 1968.)

Ipsilateral infusion of 3′,5′-cyclic AMP into the renal artery of a thyroparathyroidectomized dog results in an increased excretion of phosphate, thus mimicking the action of parathyroid hormone (Russell et al., 1968).

In mammals, at least, parathyroid hormone increases renal Ca retention. This may be due to increased tubular resorption or decreased glomerular filtration or a combination of both (Kenny and Dacke, 1975).

An interesting renal response to parathyroid hormone in mammals, which could have an important bearing on the comparative physiology of this hormone, is that of carbonic anhydrase, demonstrated by Beck et al. (1974). In crude homogeonates of renal cortex both parathyroid hormone and 3′,5′-cyclic AMP inhibited carbonic anhydrase in a dose- and time-dependent fashion. Parathyroid hormone (10 U.S.P. U ml^{-1}) and 3′,5′-cyclic AMP (0·4 mmol litre^{-1}) gave 33% and 65% inhibition of the enzyme respectively, within 20 min. Parathyroid hormone had no inhibitory effect on pure carbonic anhydrase, but 3′,5′-cyclic AMP was even more effective than in the homogeonate, indicating that the parathyroid hormone response is mediated by 3′,5′-cyclic AMP. Other hormones such as vasopressin, catecholamines and prostaglandin E which act on kidney via the 3′,5′-cyclic AMP system, had no effect on carbonic anhydrase. It was suggested that this response might account for the mechanism by which parathyroid hormone increases bicarbonate excretion. It is interesting to speculate that a similar mechanism might account for the inhibitory influence of parathyroid hormone on $CaCO_3$ secretion by the avian oviduct (see Chapter 13), a function which also involves carbonic anhydrase.

Parathyroid hormone also has effects on kidneys of sub-mammalian species including fish, which do not, apparently, possess a parathyroid hormone. Thus killifish (*F. kansae*) respond to injections of mammalian parathyroid extract (1 U.S.P. U g^{-1}) by an elevation of urinary phosphate, but not Ca excretion (Fleming, 1967).

CALCITONIN AND KIDNEY

The effect of calcitonin on renal electrolyte excretion in humans was studied by Bijvoet *et al.* (1971) who have shown that calcitonin increases the excretion of sodium, phosphate, chloride, potassium and Ca ions when administered as a constant i.v. infusion. Since these changes are also seen in hypoparathyroid subjects, they cannot be explained in terms of a secondary rise in parathyroid hormone excretion. Furthermore, they are not accompanied by an increase in the urinary 3′,5′-cyclic AMP excretion, which might be expected if parathyroid hormone were involved.

The effects on sodium, potassium and phosphate are similar to those of parathyroid hormone, but they are probably mediated at different receptor sites (Aurbach and Chase, 1976). While the natriuretic effect of the calcitonins and, in particular, the salmon hormone, is very apparent in mammals, it has not been demonstrated in sub-mammalian species. There is, however, some tenuous evidence that the hormone may affect plasma sodium and osmolality in teleost fish (see Chapter 10).

It seems that the diuretic, natriuretic and phosphaturic effects of calcitonin in rats are prevented by prior adrenalectomy, while in humans, the increased sodium and water excretion during the first 48 h of chronic treatment with the hormone, disappear thereafter, coincident with increases in plasma renin and aldosterone. These results indicate some interaction of adrenal hormones with calcitonin at the kidney level. Very large doses of cortisol may also counteract the effect of calcitonin in bone (Munson, 1976).

Vitamin D₃ Metabolites and Kidney

While the main target organs for 1,25-dihydroxycholecalciferol are undoubtedly gut and bone, it appears that physiologic doses of the metabolite may also increase renal tubular reabsorption of phosphate and Ca, an effect which is probably independent of any parathyroid hormone action on this target organ (Omdahl and De Luca, 1973).

Prolactin and Kidney

It is well known that injections of prolactin in mammals cause an increase in urinary Ca excretion (Horrobin, 1973). Whether or not this represents a direct action of the hormone on kidney, or merely reflects its hypercalcaemic activity, is not clear. It does, however, indicate that the hypercalcaemic response to prolactin is not mediated primarily by the kidney.

Endolymphatic Lime Sacs

Since endolymphatic lime sacs reach the peak of development in amphibia, it is not surprising that factors regulating their turnover have been most widely studied in this class. Virtually nothing is known of lime sac regulation, even in types such as the Dipnoi (lungfishes) or reptiles, in which the lime deposits are quite well developed.

In the amphibia, it seems that parathyroid hormone, vitamin D_3 and calcitonin all have some effect on the endolymphatic lime deposits; these effects are discussed in detail in Chapter 11.

Higher vertebrates, such as mammals, birds and some reptiles (notably Chelonia), have insignificant amounts of endolymphatic $CaCO_3$ deposits, but instead, sequester comparatively large amounts of $CaCO_3$ into their bones. Unfortunately, very little is known of the specific responses of this bone fraction, if any, to Ca regulating hormones. Acquisition of such information is clearly desirable and, in particular, it will be of interest to compare the responses of the bone carbonate fraction and those of the endolymphatic $CaCO_3$ deposits to the various Ca regulating hormones.

Skin and Scales

Skin

In fish and amphibia, the integument (skin) functions as an important organ for exchange of Ca to and from the aquatic environment. Furthermore, in fish (and some tetrapods), the bony scales also constitute an important reservoir for Ca and associated ions.

Evidence for the movement of Ca across fish skin is largely circumstantial. In experiments with *Tilapia mossambica*, Reid et al. (1959) used compartmental tanks to separate anterior and posterior portions of the fish. In this situation, the posterior portion of the integument was able to accumulate ^{45}Ca from the water independently of mouth or gills. Other workers (Mashiko and Jozuka, 1964) consider that ^{45}Ca may be taken up selectively by the fins and gills of fish, since enclosure of the caudal fin by a rubber sac containing ^{45}Ca resulted in a particularly large uptake of the isotope by that organ. These workers also concluded that fins and gills are important sites for Ca excretion in fish.

As yet, there are no reports concerning effects of hormones on either Ca uptake or excretion via the fish integument, except those specifically concerning the gills (see below).

Amphibian skin may also be involved in Ca translocation. Frog skin *in vitro* has been studied by Watlington et al. (1968), who observed both influxes and effluxes of radiocalcium across this tissue. Addition of bovine parathyroid hormone (0·35 U.S.P. U ml^{-1}) to the chamber, elicited a significant increase in ^{45}Ca influx of about 60% more than the control value, while radiocalcium efflux was decreased by 75% compared with controls. Both effects occurred within 1 h of parathyroid hormone treatment. Electrical studies in this system revealed no significant changes in either intermittent open circuit potential difference or in the short circuit current required to abolish this potential difference. The authors concluded that at least two mechanisms are involved in the frog skin response to parathyroid hormone, one involving non-electrogenic active transport of Ca, the other involving a decrease in Ca permeability.

At present, there do not appear to be any published studies of calcitonin, vitamin D_3 metabolites or any other Ca regulatory hormones in the frog skin Ca transportation preparation.

SCALES

There is considerable evidence in the literature to suggest that the Ca deposits of fish scales are labile (Simkiss, 1974). Evidence for the hormonal modulation of Ca turnover in these scales comes from a variety of sources. Roach (*Rutilus rutilus*), when maintained on rachitogenic diets, showed evidence of scale resorption starting at the anterolateral corners, spreading to the cranial edge and attacking the exposed region. This phenomenon also occurs in wild roach populations during the reproductive season of mid-summer (Wallin, 1957). Similarly, Garrod and Newell (1958) demonstrated a decline of around 20% in the Ca content of scales of female *Tilapia esculenta* at a time when the ovaries were actively developing prior to the onset of reproduction, while recently, Mugiya and Watabe (1977) demonstrated that a single injection of oestrogen causes scale resorption in goldfish and killifish (see Chapter 10).

Other results, from Fleming and his group, indicated direct effects for both calcitonin and the pituitary hormones on Ca turnover in scales of the acellular boned teleost, *F. kansae*. These data are reviewed in more detail in Chapter 10.

Gills

Fish gills are well known as targets for hormonal action. They are involved in the uptake and excretion of water and electrolytes such as sodium, potassium, ammonium, protons, chloride and bicarbonate, and in this respect respond to hormones such as vasotocin, cortisol, prolactin and the catecholamines (Bentley, 1976). Some of these hormones i.e. catecholamines and the neurohypophysial peptides, alter blood flow to the gills and are, therefore, likely to have general effects on hydromineral exchange, including that of Ca and associated ions. The gills are essentially an extension of the integument and probably reflect the functional characteristics of skin with respect to exchange of electrolytes. As mentioned previously, Mashiko and Jozuka (1964), consider the gills, along with fins, to be particularly important in Ca exchanges between the environment and body fluids. A number of recent reports have suggested that Ca may be actively transported across the teleostean gill and that this transport may be modulated by hormones.

Attention has begun to focus on the role of Ca activated ATPases in fish gills. Recently Ma *et al.* (1974) described a ouabain insensitive, sodium and potassium independent, ATPase from the gills of rainbow trout (*Salmo gairdneri*) which was preferentially activated by Ca ions. A similar enzyme exists in gills of the American eel (*Anguilla rostrata*) (Fenwick, 1976). In freshwater adapted eels, removal of the corpuscles of Stannius resulted in a doubling of gill tissue Ca^{2+}-ATPase activity (in terms of moles Pi released per g per wet gill filament per h) from $2·6 \pm 0·2$ in controls to $4·2 \pm 0·3$ in Stanniectomized fish. The kinetic properties of the enzyme from the two groups were similar, indicating that the total number of enzyme molecules was probably increased by the treatment. Stanniectomy also resulted in an increased net uptake of ^{45}Ca in individually perfused isolated eel gills after the same post-operative interval (Fenwick and So, 1974), leading Fenwick (1976) to interpret his results as evidence for a role of the Ca^{2+}-ATPase as an energy source for the development of post-Stanniectomy hypercalcaemia. While this cannot be taken as definitive evidence for a relationship between Ca transport across gills with a specific Ca^{2+}-ATPase, there is considerable evidence from other tissues such as red blood cells and sarcoplasmic membranes, that the activity of Na^+/K^+-ATPase and Ca^{2+}-ATPases can vary with the transport of their respective ions (Fenwick, 1976). This would appear to be the first direct evidence for an effect of a Ca regulating hormone (hypocalcin) on fish gills, its activity being of an inhibitory nature. There also appears to be a specific Ca binding protein in teleost fish gills with properties similar to that found in the intestine (Chartier, 1973). In eels (*A. anguilla*) maintained in fresh water, removal of Stannius corpuscles resulted, within 2 weeks, in an increase in ^{45}Ca binding activity of about 50% (on a whole body basis), although in terms of gill wet wt, the smaller increase was not significant (Chartier et al., 1977). These authors have also detailed the cytological changes in terms of chloride cell proliferation and hypertrophy which occur following Stanniectomy. Apparently, none of these responses occur when eels are maintained in Ca free water. Changes in chloride cell cytology are commonly correlated with general alterations of hydromineral regulation in fish (Maetz, 1974), which are themselves profoundly influenced by acalcic conditions, but it is possible that these changes may also be involved in phospho-calcium transport.

Recently, Ca translocation in isolated *in situ* gills of *A. anguilla* following either ultimobranchial gland "removal" or calcitonin treatment, has been investigated by Peignoux-Deville et al. (1978). Salmon calcitonin at a dosage of 20 ng ml^{-1} reduced Ca influx and increased efflux (see Fig. 27). Removal of ultimobranchial tissue by electrocautery also caused a reduction in Ca influx

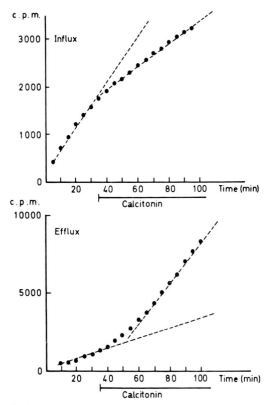

FIG. 27. Effect of salmon calcitonin (CT) perfusion (20 ng ml^{-1}) on ^{45}Ca influx and efflux in the perfused eel gill *in situ*. (●-●-●) represent flux values following CT treatment (- - -) represent control values. (From Peignoux-Deville et al., 1978.)

of more than twofold when compared with control animals, but Ca efflux was not affected by this procedure. The authors suggested that ultimobranchialectomy may have activated the Stannius corpuscles which, in turn, reduced Ca influx, and certainly this would be consistent with the data from Fenwick and So (1974).

Oviduct

There is considerable evidence that in viviparous mammals, the placenta can actively transfer Ca ions from maternal to foetal side, although it is not yet clear whether this potential target organ is sensitive to hormones which regulate Ca (Simkiss, 1967). Most vertebrate species lay eggs which are released into the environment and the developing embryos within such eggs probably obtain sufficient Ca for their needs from the yolk. As mentioned in Chapter 5, the yolk precursors formed in the liver of sub-mammalian species, from bony fish to birds, bind Ca avidly.

Of especial interest to this part of the discussion are the cleidoic eggs of reptiles and birds which relate to a method of reproduction differing from that of amphibia and other lower vertebrates in that no primitive larval stage is produced at hatching. Instead, the hatching is morphologically similar to the adult type and has ossified bones. The need for extra Ca which this method of reproduction entails, is met, in part, from Ca reserves of the yolk. In certain reptiles, for example some snakes and lizards, the egg is retained in the oviduct until hatching, these animals thus being ovoviviparous. Other species have developed true viviparity, while some species of snakes and lizards are oviparous and their large eggs are enclosed in a leathery shell. The shells of crocodile and turtle eggs are thick and heavily impregnated with $CaCO_3$ and this latter adaptation is also found in birds' eggs. The hard, calcified eggshell obviously has a major protective function, but also acts as a source of Ca for the developing embryo (Simkiss, 1967).

The process by which Ca is secreted by the oviduct to form eggshells in hens, has attracted considerable attention over the last 30 or 40 years because of its commercial implications. Evidently $CaCO_3$ secretion by the oviduct is a thermodynamically active process (Schraer and Schraer, 1965, Schraer et al., 1965). The quail oviduct contains a Ca^{2+}/Mg^{2+} activated ATPase, located in the microsomal fraction of the shell gland mucosa. The mean activity of this enzyme in the gland of ovulating hens is about one and a half times that found in pre-calcifying birds, while it is three times that of immature birds (Pike and Alvarado, 1975).

Only recently has indirect evidence been published to indicate that the process of mineral secretion by the oviduct might be modulated by hormones. Evidence from Corradino et al. (1968) suggests that the avian oviduct contains a Ca binding protein similar to that of the gut and which may be sensitive to changes in vitamin D_3 status. This, however, is disputed by Bar and Hurwitz

(1973), who were unable to alter the oviducal binding protein activity during vitamin D_3 deprivation. The Ca binding activity of egg-laying Japanese quail oviduct has been demonstrated as being almost seven times greater than in non-laying hens, suggesting the protein does play a role in translocation of Ca from blood to eggshell in the laying bird.

Indirect evidence for an inhibitory role of parathyroid hormone on the process of eggshell secretion by the avian oviduct has recently been demonstrated by Dacke (1976); this data is considered in detail in Chapter 13.

7. Calcium Regulation in Fish—General Considerations

The term "fish" encompasses several major classes of aquatic vertebrates with gills. These are often considered together as the superclass Pisces, which may or may not include species with integumentary outgrowths (i.e. fins). The classification into superclass Pisces belies the heterogeneity of this large group of vertebrates and it is much more meaningful to consider them as separate classes comprising the Agnatha (primitive, jawless vertebrates), Chondrichthyes (sharks and rays) and Osteichthyes (modern bony fish). The Agnatha, which are represented by few extant species, are also considered to include the extinct Ostracodermi, a group in which rudimentary bone was present. A more advanced fossil group, the Placodermi, is often considered by taxonomists to form a separate class. For detailed relationships of the superclass Pisces, the reader is referred to Fig. 1. Virtually the only thing that these vertebrate classes have in common, is that they inhabit aqueous environments. As far as their calcium (Ca) regulation is concerned, fish are probably even more diverse than the tetrapods. Indeed, Simkiss (1974) considers that "fish were the first and in many ways the last of the great modifiers of calcium metabolism in the vertebrates".

Any attempt in the next few chapters to subdivide our consideration of fish Ca regulation must be arbitrary. On the one hand, they may be considered along phylogenic lines, on the other, on the basis of presence, or absence, of calcified bone. A third method of division could involve a consideration of the aquatic environment, fresh water versus sea or brackish water. The electrolyte physiology of fish living in these different environments will vary considerably, and of special interest is electrolyte metabolism in euryhaline fish (i.e. those migrating between fresh and saline waters for breeding or other purposes).

Before discussing Ca regulation in fish along phylogenic lines, we should first consider some general aspects concerning regulation of the electrolyte in these animals. First, what are the most important environmental pressures facing fish? Second, what major systems are used by fish in their regulation of Ca?

CALCIUM REGULATION IN FISH—GENERAL CONSIDERATIONS 91

The Environment

Fish live in many diverse environments including fresh water, the sea or estuarine waters which represent a brackish half-way house in terms of salinity between the sea and river waters. Fresh water can be very variable in Ca content, ranging from soft to hard waters. Sea water does not represent the upper

TABLE XV
Concentrations of Ca and total ions in the plasma of various groups of fish

	Plasma Ca (mmol litre^{-1})	Total ions (mmol litre^{-1})
Agnatha		
Order Cyclostamata		
Petromyzon tridentata (lamprey)	2·8 ± 0·3	186 ± 41
Polistotrema stoutii (hagfish)	5·4 ± 0·1	1026 ± 55
Chondrichthyes		
Subclass Elasmobranchii		
Heterodontus francisa (horn shark)	5·0 ± 0·2	490 ± 20
Carcharhinus leucas (bull shark)	4·5 ± 0·5	483
Subclass Holocephali		
Hydrolagus colliei (ratfish)	4·8 ± 0·3	558 ± 15
Osteichthyes		
Subclass Actinopterygii		
Chondrostei		
Acipenser transmontanus (sturgeon)	1·8 ± 0·4	256
Polypterus weeksi (bichir)	2·3 ± 0·1	201
Polyodon spathula	2·2 ± 0·1	250
Holostei		
Lepisosteus platystomus (gar pike)	2·6 ± 0·2	318
Amia calva (bowfin)	2·9 ± 0·2	275
Teleostei		
Oncorhynchus tschawytscha (Pacific salmon)	2·9 ± 0·2	281 ± 26
Coregonus clupeoides	2·7	275
Megalops atlantica	2·5 ± 0·1	267 ± 35
Paralabrax clathratus	3·2 ± 0·0	353 ± 24
Subclass Choanichthyes		
Coelacanthini		
Latimeria chalumnae (coelacanth)	3·4	c.454
Dipnoi		
Neoceratodus forsteri (Australian lungfish)	1·6	—
Protopterus annectens (African lungfish)	1·8 ± 0·2	210
Lepidosiren paradoxa (S. American lungfish)	2·20 ± 0·3	224
Sea water (Pacific)	10·0 ± 0·1	1168 ± 60
"Fresh water" (Lake Huron)	0·9 ± 0·1	5 ± 2

(From Simkiss, 1974.)

extreme of salinity inhabited by fish; sometimes, inland streams or lakes contain brackish waters with concentrations of salts even higher than in the sea. Most fish live in waters of fairly specific salt content, whether fresh water or sea water, and when placed in the alternative environment these fish usually succumb to the osmotic stress and die; such fish are termed "stenohaline". Relatively few fish, mostly representatives of the class Teleostei, are able to inhabit waters of widely varying salinities. This may occur at any time in species such as estuarine flounder (*Platichthys flesus*) or during migration preparatory to breeding, as in salmon and some cyclostomes, for example, the lamprey (*Lampetra fluviatilis*) (Bentley, 1976). These fish, which are termed "euryhaline", have been studied mainly with regard to regulation of water and ions such as sodium and chloride. They can also encounter widely differing concentrations of divalent ions such as Ca and magnesium in their peregrinations, but regulation of these, particularly the latter ion (for reasons outlined in Chapter 14), has received scant attention.

The differing concentrations of salts in the aquatic environment are shown in Tables I and III, while the concentrations of Ca in the plasma of a variety of fish are shown in Table XV. It is clear from these that even the most primitive living fish, the cyclostomes, regulate their plasma Ca levels, while the Osteichthyes have plasma Ca levels which are quite similar to those found in mammals.

Open and Closed Calcium Systems

There are two basic systems by which vertebrates regulate their plasma Ca levels. These are (1) open systems, in which the animal contains little or no internal reservoir of Ca and, therefore, regulates Ca by direct exchange with the environment, and (2) the so-called closed system in which most of the plasma Ca can be recycled within the animal by referral to an internal Ca reservoir. Open systems are found mainly in cyclostomes and chondrichthyan fish, which are not considered to have any major Ca reservoirs (Urist, 1964, 1976b). Instead, these fish exchange Ca directly with the environment via gut, gills, skin and kidney, and, since these species live predominantly in the sea, which contains large quantities of Ca, the sea water can be considered as an external Ca reservoir. Of course, in this situation, the fish has a problem of an overadequate supply of Ca and must take steps to exclude it. Some species of cyclostomes and chondrichthyans inhabit freshwater environments and these must be able to concentrate Ca from the environment.

Closed systems of Ca regulation are found in bony fish and tetrapods. Presumably the most primitive exinct bony fish, the Placodermi and Ostracodermi, had closed systems, since they formed bone or dentine.

The basic mechanisms of open and closed Ca regulating systems in fish are summarized in Fig. 28.

CALCIUM REGULATION IN FISH—GENERAL CONSIDERATIONS 93

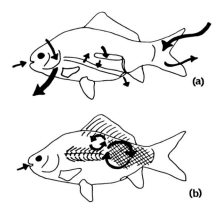

FIG. 28. (a) A fish considered as an "open system". Ca mainly enters and leaves via the fins and gills. A small amount of Ca is absorbed via the intestine and the kidneys exert some control over Ca loss. In its most extreme form this concept considers that the surrounding water acts as an extension of the body fluids. (b) A fish considered as a "closed system". Ca is stored in the scales and skeleton and continually recycled in and out of the structures so as to maintain a steady-state in the animal. (a) and (b) represent extreme forms and most teleosts probably have both systems which are clearly not mutually exclusive. (From Simkiss, 1974.)

In the following three chapters, fish Ca regulation is reviewed along phylogenic lines. Only extant classes are considered since we cannot be sure which of the control systems and target organs, other than mineralized tissues, were present and functional in extinct species.

8. Calcium Regulation in the Agnatha

Target Tissues and Blood Electrolyte Levels

Cyclostome fish are the living representatives of the class Agnatha or jawless vertebrates. These are the most primitive living vertebrates and are highly degenerate in terms of calcified tissues, having completely lost any ancestral bone or scales, if these tissues were ever present. Cyclostome cartilage appears to be primitive in that the cells are large and of the invertebrate type. Furthermore, there is little intercellular basophilic matrix present (Urist, 1962). The only calcified tissues found in cyclostomes are small apatite otoliths in the inner ear. However, these animals probably evolved directly from primitive vertebrates, the Ostracoderms (see Fig. 1) in which rudimentary bony tissue was present.

Cyclostome fish are further subdivided into two orders, (1) Myxiniformes (hagfish) and (2) Petromyzontiformes (lampreys). Hagfish are scavengers, inhabiting an exclusively marine environment, but lampreys invariably breed in fresh water. Some lamprey species spend their entire life in fresh water and are generally parasitic for part of their life cycles.

Plasma calcium (Ca) levels have been reported for several species of cyclostomes (Holmes and Donaldson, 1969, Urist, 1976b). The ion is always present at a lower concentration in plasma of seawater species than in sea water itself, but higher than the water concentration in freshwater species. Plasma Ca concentrations range from 6·3 mmol litre^{-1} in the marine *Myxine glutinose* to 2·0 mmol litre^{-1} in the marine *Lampetra fluviatilis*, while in freshwater species they range from 1·8 mmol litre^{-1} in metamorphosed *Petromyzon marinus* to 2·4 mmol litre^{-1} in the ammocoete larva of the same species. It is clear from these figures that the plasma Ca levels are much more variable in cyclostomes than in higher vertebrates and they reflect, to some extent, Ca levels in the environment. Urist (1963) transferred metamorphosed potodramous lampreys (*P. marinus*) from fresh water into artificial sea water and observed changes in the serum concentration of electrolytes after 2 and 4 h of immersion. In 4 h, the serum Ca concentration increased from 1·8–3·5 mmol litre^{-1} while serum inorganic phosphorus rose from 2·7–4·3 mmol litre^{-1}. He also transferred these fish from fresh water to Ca deficient fresh water for 2 h which resulted in a small drop in serum Ca concentration from 2·6–2·2 mmol litre^{-1} and an increase in serum phosphate concentration from 3·0–4·8 mmol litre^{-1}. The phosphate rise is rather similar to the response in bony vertebrates during hypocalcaemia. These findings, particularly the marked stability of serum Ca levels during transfer to Ca deficient fresh water, indicate that lampreys are able to regulate Ca despite the absence of a skeletal reservoir for the electrolyte. Urist (1976b) considers that the gills and mucosal epithelium are important in Ca regulation in cyclostomes. Presumably the gut and kidney are important organs for the regulation of divalent ions as in higher fish. In hagfish, magnesium and phosphate are secreted into the glomerular filtrate by the mesonephric duct cells and appear in the urine at higher concentrations than in the plasma (Munz and McFarland, 1964). Calcium, however, is reabsorbed by the nephra, at least in marine cyclostomes (Hickman and Trump, 1969) who also demonstrated that the liver may act as a major excretory pathway for Ca and magnesium ions via the gall bladder. It remains to be seen if these organs can transport Ca as they do in some higher vertebrates. Certainly the gut, gills and skin of cyclostomes seem to be primitive in that there is no direct evidence for active transport of sodium across their surface membranes as there is in teleost fish (McFarland and Munz, 1965). In contrast to its low sodium content, the epithelial slime has a high Ca, magnesium and potassium content which suggests that Ca can be secreted across the epithelium, at least in this form (McFarland and Munz, 1965).

Hormones

Of the three major Ca regulating hormones in higher vertebrates, only vitamin D_3 appears to have been tested for activity in cyclostomes. Neither oral nor intramuscular administration of the native vitamin has any effect upon plasma Ca levels in these fish (Urist, 1974). As mentioned previously (Chapter 5), the liver of cyclostomes contains little, if any, vitamin D_3. Unfortunately, the liver and kidneys of these animals have yet to be tested for the presence of vitamin D_3 hydroxylase activity. Calcitonin, parathyroid hormone and hypocalcin are all lacking in cyclostomes. The hypercalcaemic pituitary hormone, prolactin is possibly absent in this subclass, while there is some doubt as to whether or not corticosteroids such as cortisol are present (Bentley, 1976).

It appears, then, that the cyclostomes are able to regulate and conserve Ca, but perhaps not as well as higher vertebrates. This is accomplished in an open system using primitive membranes which incorporate mechanisms for cycling Ca, phosphate and other ions in large volumes of water of the hydrosphere. None of the major hormone systems involved in Ca regulation in higher vertebrates, apart, possibly, from corticotrophin, cortisol and prolactin, appear to be present in cyclostomes. It remains to be seen if other hormones can modulate the intake and excretion of Ca in these species.

In conclusion, it can be stated that cyclostomes manifest the most primitive of Ca regulatory systems to be found in vertebrates.

9. Calcium Regulation in the Chondrichthyes

The Chondrichthyes are a class of fish comprising the sharks and rays (subclass Elasmobranchii) and the chimaeras or ratfish (subclass Holocephali). The latter subclass is represented by a few sparsely distributed species. Chondrichthyian fish probably evolved from the primitive fossil Placodermi in which bone was present (see Fig. 1). Vestiges of bone are represented by the dermal denticles which are covered with enameloid in some species of Elasmobranchii. In the Holocephali, even these vestigial scales are absent and in common with cyclostomes, only a naked mucosal epithelium is present on the exterior surface. The main skeletal material of the Chondrichthyes is calcified cartilage, although in the Holocephali, this may be present only in scanty patches (Urist, 1976b). Other calcified tissues in the Chondrichthyes are small otoliths of $CaCO_3$ in the inner ear. The Elasmobranchii are predominantly seawater dwellers although several species inhabit fresh water. The Holocephali are entirely seawater fish.

Blood Electrolyte Levels

Plasma Ca levels have been reported for several species of chondrichthyans (Holmes and Donaldson, 1969; Urist, 1976b). Marine species, like the cyclostomes, have plasma osmolalities which are often similar to, or a little lower than in the sea water itself. These are accounted for by large quantities of urea, since the Chondrichthyes, in common with all tetrapods, but not with other fish (except Coelacanths), retain an ornithine cycle by means of which ammonia is converted to urea in the liver. Urea is relatively non-toxic but osmotically active, so these fish are able to maintain high plasma osmolalities but without excessively high levels of physiologically active electrolytes, or the high levels of metabolic expenditure necessary to regulate them, at least in sea water. Plasma Ca levels in this class are quite variable, but for seawater species they are generally about half of that in the surrounding environment. Plasma Ca levels in the Chondrichthyes range from 11·6 mmol litre^{-1} in a marine species *Dasyates americana* to 3·3 mmol litre^{-1} in the marine *Scyllium canicula*. Most seawater species have typical values around 4 to 5 mmol litre^{-1}.

In freshwater species, the plasma Ca levels range from 3·0 mmol litre^{-1} in

Carcharhinus leucas nicaraguensis to 4·8 mmol litre^{-1} in the ratfish (*Hydrolagus collie*) (Holmes and Donaldson, 1969). Plasma Ca levels are generally higher in seawater compared with freshwater species. There do not appear to be any studies of the type carried out in cyclostomes or higher bony fish, concerning changes in plasma Ca levels which might be induced by substitution of sea water with fresh or Ca deficient water.

Urist (1961) has examined the serum Ca concentration in several species of chondrichthyan, both male and female, mature and immature, gravid and non-gravid. He found that while Ca concentrations are generally high, there do not appear to be any sex differences comparable to those in bony fish or tetrapods. This might be explained by the low albumen content of chondrichthyan plasma (Urist, 1976b). While the plasma proteins comprise 1–3 g 100 ml^{-1} of globulin, they contain only a trace of albumen. In the Chondrichthyes this has an important effect on the protein binding of electrolytes, with 80% of the total plasma Ca in these species being non-protein bound; of this, about 80% appears to be in the ionized form as estimated by a variety of methods.

Target Tissues

In terms of target organs, the chondrichthyans are similar to cyclostomes in that they have a series of membranes, the gut, gills, skin and nephron, which influence the exchange of Ca between the body fluids and environment. In addition, there are rectal glands capable of secreting large quantities of salt in marine habitats (Urist, 1976b). It is not yet clear, however, if this latter organ has any role in Ca homeostasis.

There are no major Ca reservoirs other than calcified cartilage. While the Ca and phosphate reserves of this tissue may be available by direct exchange for short-term needs, there is little evidence that they can provide a controlled supply of Ca and phosphate for long-term needs, since in order to fulfil this function, the tissue would need to have a rather limited capacity for exchange with the body fluids (see Chapter 4). The Ca regulatory system in these chondrichthyans can thus be considered as an example of the open type.

Hormonal Influences

Chondrichthyans differ from cyclostomes in that they have a larger complement of hormones which may be involved in Ca regulation; these include cortisol or its adrenal homologues and pituitary prolactin (Bentley, 1976). Calcitonin is also present but vitamin D_3 and its metabolites are probably absent (see Chapter 5). The corpuscles of Stannius and hence hypocalcin, are absent in chondrichthyans, as is parathyroid hormone. In this vertebrate class, then, we see the first definite appearance of some of the major Ca regulating hormones found in higher vertebrates. Whether or not these hormones have a Ca regulatory function in the Chondrichthyes is a different matter; there is very little evidence for any of the hormones mentioned above having such a function in this class.

Calcitonin from the ultimobranchial glands of sharks has a hypocalcaemic effect when injected into mammals (Copp et al., 1967a). However, injections of either mammalian or homologous calcitonins had no effect on serum Ca levels in a variety of elasmobranch species, and neither did the hormone influence renal function (Hayslett et al., 1971). The urine Ca concentration in a marine shark, *Squalus acanthias*, is at 3·0 mmol litre^{-1}, about 86% of the plasma Ca concentration. This value is similar to that found in cyclostomes, but much lower than that in teleosts (Hickman and Trump, 1969), which suggests that the kidney is not an important organ for the maintenance of plasma Ca levels in these two lower classes of vertebrates, at least when they are in a seawater environment.

Since bone is a primary site for calcitonin action in higher vertebrates and as chondrichthyans lack this tissue, it is not surprising that the hormone has no hypocalcaemic effect in these fish. There are three questions which are posed by the presence of calcitonin in chondrichthyans. (1) Does the hormone have any Ca regulatory function in these fish, and if so, what target organs and responses are involved? (2) Does the hormone have a function other than Ca regulation, such as that of phosphate, sodium, magnesium or water? (3) Is the hormone vestigial in the Chondrichthyes, having evolved as a functional hormone in their bony ancestors? Only the first two questions can be answered directly by research, while the third possibility would only become viable after exhaustive research has failed to demonstrate any response to calcitonin in the class Chondrichthyes.

Administration of vitamin D_3 at high dose levels to sharks and rays had no effect on their plasma Ca concentration (Urist, 1972). Furthermore, the livers of these species contain only negligible amounts of the vitamin (Urist, 1976b).

Other vertebrate Ca regulating hormones such as hypocalcin, prolactin, cortisol and parathyroid hormone have not, apparently, been tested for calcaemic activity in the Chondrichthyes.

Injections of oestrogens gave slight (11%) hypercalcaemic responses in dogfish (*Scyliorhinus caniculus*) according to Woodhead (1969), but in other sharks (*Triakis semifasciata* and *Heterodontus francisci*), these substances had no effect on plasma Ca levels. Since there are no sex differences in plasma Ca levels of chondrichthyan species (Urist, 1961), it is doubtful that the small response noted by Woodhead (1969) has any physiological significance.

10. Calcium Regulation in the Osteichthyes

The Osteichthyes (bony fish) are the highest of fish classes and comprise two subclasses, the ray-finned Actinopterygii and fleshy-finned Sarcopterygii (or Choanichthyes). These are the most successful and widespread of the fish classes, inhabiting a variety of environments ranging from soft, fresh waters through highly saline, brackish waters to sea water. Some species even breathe air, having evolved rudimentary lungs, and are thus able to withstand some degree of desiccation. The Actinopterygii are represented by species ranging from those with almost completely cartilaginous endoskeletons (the Chondrostei) to those with highly ossified endoskeletons (the Holostei). The major order in this subclass, the Teleostei, often have specialized endoskeletons containing a considerable quantity of acellular bone. These are the most abundant and successful of all fish and are, therefore, economically important as a food source, which is reflected by the huge amount of literature concerning Teleostean biology. The Sarcopterygii comprise the airbreathing lungfishes (Dipnoi) and the rare coelacanths (Crossopterygii). Both these orders, although sparsely represented, constitute important evolutionary links between fish and tetrapods.

Blood Calcium Levels

Plasma calcium (Ca) levels have been reported for several species of osteichthyan fish and these are generally lower than those reported for members of classes Agnatha or Chondrichthyes. Typical values for seawater species range from 1·5 mmol litre^{-1} in the chondrostean *Acipenser oxyrhynchus* to 8·3 mmol litre^{-1} in the teleost *Spheroides maculatus*, but most seawater species have typical values around 2–4 mmol litre^{-1}. Freshwater species have typical plasma Ca values ranging from 1·25 mmol litre^{-1} in *Acipenser fulvescens* to 4·9 mmol litre^{-1} in *Micropterus dolomieue*. Most freshwater species have values between 2 and 4 mmol litre^{-1} but they are generally slightly lower than for seawater species.

When euryhaline species of osteichthyan fish are transferred from fresh water to sea water or vice versa, accompanying changes in the plasma Ca levels occur. For example, the killifish (*Fundulus kansae*) has a plasma Ca level of 2·5 mmol litre^{-1} in the freshwater condition. If it is transferred to sea water, the plasma Ca level rises to 4·6 mmol litre^{-1} within 3 days; after 20 days in sea water, however, the plasma Ca level declines to 3·0 mmol litre^{-1}, this change being similar to the plasma sodium change which occurs in the same species during the same treatment (Stanley and Fleming, 1964). Similarly, the eel (*Anguilla anguilla*) in the "yellow" condition, shows a rise of plasma Ca concentration from 2·31 mmol litre^{-1} in fresh water to 2·75 mmol litre^{-1} following 4 weeks adaptation to sea water. Silver eels of the same species undergoing similar treatment exhibit only slight changes in plasma Ca level (2·29 to 2·37 mmol litre^{-1}) according to Chan *et al.* (1967). No data were given for the acute plasma Ca responses in these experiments. The same authors reported plasma Ca concentrations for eels maintained in distilled water, a medium in which this species survives particularly well in comparison with other fish. Paradoxically, plasma Ca in these fish rose, after 4 weeks of this treatment, to levels of 2·90 mmol litre^{-1} for "yellow" eels and 2·89 mmol litre^{-1} for "silver" eels; this was despite the fact that the plasma sodium concentration showed a significant fall. The explanation of the plasma Ca response is not clear, but presumably it reflects an increased mobilization from internal Ca reservoirs. In these experiments, the plasma inorganic phosphorus level also rose from 1·30 to 1·86 mmol litre^{-1} in "yellow" eels on transfer from fresh to distilled water, while it rose to 2·02 mmol litre^{-1} following adaptation to sea water. The plasma phosphate rise in the distilled water experiment indicates increased mobilization of bone or scale mineral. The terms "yellow" and "silver" eel refer to different stages of the life cycle of this species; a major part of the life cycle is spent as a "yellow" eel in fresh water. When the fish becomes sexually differentiated prior to its catadromous migration, it stops feeding and the ventral surface turns a silvery white colour.

The requirements of Ca (and other electrolyte) regulation are met, in bony fish, by their transfer directly from the environment to body fluids or vice versa via the gills, gut, kidney, skin and fin surfaces. In addition, the bony endoskeleton and scales are important reservoirs for Ca, phosphate and other ions. Of interest is the fact that bony fish, in contrast to the lower (non-bony) vertebrates, possess a small proportion of extracellular fluid relative to that of intracellular fluid (Thorson, 1961). It seems that the extracellular fluid is used more efficiently by bony vertebrates, perhaps by virtue of the presence of the bone mineral reservoir.

The role of hormones in regulating these systems in teleost fish has been studied in very great detail and it is unfortunate that other orders of fish have not received the same attention.

Calcitonin and the Ultimobranchials

Since 1954, it has been recognized that the ultimobranchial body plays some role in teleost Ca regulation. In the often cited, but curious work of Rasquin and Rosenbloom (1954), Mexican cave fish, *Astynax mexicanus*, were maintained under conditions of complete darkness for two years. This treatment resulted in distortions of the vertebral columns of these fish, indicating a loss of bone mineral. Furthermore, calculi were deposited in the kidneys, while histological studies indicated hyperplastic responses by the ultimobranchial bodies, accompanied by thyroid, interrenal and, occasionally, gonadal atrophy. The authors suggested that a parathyroid-like function could be ascribed to the ultimobranchial body and that its overactivity could give rise to the excessive bone resorption. We now know that the ultimobranchial is the site of calcitonin production and, therefore, its hyperplastic response probably reflects a high rate of calcitonin synthesis and secretion in the face of a Ca stress.

More recently, Pang (1971a) has suggested that rather than absence of light being the cause of Ca stress in the above experiments, it was the result of a dietary deficiency of vitamin C. His studies of the effect of darkness and vitamin C deficiency on Ca metabolism in seawater adapted killifish (*Fundulus heteroclitus*), showed that while adult fish did not respond to complete darkness, juvenile fish showed a 20% hypocalcaemic response after 15 weeks and signs of skeletal deformity after 28 weeks. These changes were prevented by dietary supplementation with vitamin C (100 mg 100 g^{-1} food). It is not clear from these experiments, or those of Rasquin and Rosenbloom (1954), whether the hypocalcaemic response was a result of ultimobranchial overactivity or a cause of it. In view of the well-known effect of calcitonin in blockading bone resorption in higher vertebrates, and probably also in teleost fish (see below), it seems unlikely that the bone resorption picture in the aforementioned experiments could be a result of calcitonin hypersecretion.

Since the demonstration by Copp *et al.* (1967a, b) that the ultimobranchial bodies of sub-mammalian vertebrates represent the source of calcitonin, considerable interest has been shown in the function of this organ in fish. At the present time, however, there is still much confusion concerning its role in these species. Three main types of investigation have so far been carried out in fish, (1) effects of injected calcitonin (2) effects of ultimobranchial extirpation and (3) studies on the physiological responses of circulating calcitonin levels.

CALCITONIN INJECTION

There is considerable disagreement on the question of whether or not calcitonin has any hypocalcaemic effects when injected into teleost fish. Hog calcitonin was first reported by Pang and Pickford (1967), to have no hypocalcaemic effects on the acellular boned killifish (*F. heteroclitus*), but in the cellular boned catfish (*Ictalurus melas*), the hormone was reported to have hypocalcaemic and hypophosphataemic effects (Louw *et al.*, 1967). In the latter investigation, however, the four control fish had rather high and variable plasma Ca levels (4·0 ± 1·0 mmol litre^{-1}) and there is some question as to whether or not this group contained females with mature ovaries and, hence, high levels of protein bound plasma Ca. Furthermore, in juvenile channel catfish (*Ictalurus punctatus*), no hypocalcaemic response was found up to 8 h after injection of either porcine or salmon calcitonin (Pang and Grant, cited by Pang, 1971). In these experiments the fish were maintained in either fresh water (0·15 mmol litre^{-1} Ca) or Ca enriched fresh water (7 mmol litre^{-1} Ca). Unfortunately, the doses of hormone used in these experiments were not reported. The eel (*A. anguilla*), which also has cellular bone, was reported to respond to mammalian calcitonin by Chan *et al.* (1968). These authors demonstrated a hypocalcaemic response of around 20% within 5 to 6 h of injection of 50 MRC mU 100 g^{-1} body wt of porcine calcitonin, while phosphate levels increased. The experiments were carried out in freshwater adapted fish and apparently the responses disappeared in fish which had previously had their corpuscles of Stannius removed surgically. Other workers have failed to repeat these findings with respect to the hypocalcaemic responses to injected calcitonin in either European eels (Dacke, 1972) or American eels (Hayslett *et al.*, 1971). The experiments of Dacke were carried out in the same laboratory with presumably the same freshwater supply and stock of fish originally used by Chan *et al.* (1968). In these experiments, however, doses of up to 200 MRC mU 100 g^{-1} body wt of either partially or highly purified porcine or synthetic salmon calcitonin were used. Elsewhere, Pang (1971b) has obtained hypocalcaemic and hyperphosphataemic responses in the American eel (*Anguilla rostrata*). Other workers who failed to obtain hypocalcaemic responses to calcitonin in teleost fish include Orimo *et al.* (1971), Lopez *et al.* (1971) and Copp *et al.* (1972). In these latter experiments (Lopez *et al.*, 1971 and Copp *et al.*, 1972) the effects of salmon calcitonin on both blood and urine Ca levels in salmon were reported, but these were negative. However, Lopez *et al.* (1971) did find a hypocalcaemic response in trout (*Salmo gairdnerii*) if maintained in acalcic (but not normal) fresh water, suggesting that the environmental Ca supply may be an important factor in the response to calcitonin, possibly by lowering the endogenous level of the hormone.

Orimo *et al.* (1971, 1972) have reported that eel calcitonin injected into freshwater eels (5 MRC U per fish) results in a decrease in serum osmolality, sodium and chloride (see Fig. 29). Apparently, in these studies, no proper control experiments were performed and it is possible that haemodilution could have produced the observed changes in serum hydromineral values. In killifish (*F. heteroclitus*), injection of either mammalian or salmon calcitonin has been reported to produce a hypochloraemic response although this effect was only evident in fish receiving 14 or more injections of the hormone (Pang, 1973). The hypochloraemia was most marked in fish maintained in Ca deficient sea water. Plasma chloride fell from control levels of 167 mmol litre^{-1} to 136 mmol litre^{-1}, a fall of 20%. Neither plasma sodium nor Ca exhibited any significant changes in these experiments, but unfortunately, once again the hormonal dose levels were not reported, thereby making it impossible to interpret these results quantitatively.

Dacke (1975) has studied the effect of salmon calcitonin on efflux of acutely injected ^{22}Na and ^{45}Ca labels from freshwater adapted eels (*A. anguilla*). Total efflux of label was measured in these experiments and it was assumed that whereas most of the sodium efflux occurs across gills, that of Ca occurs partly via gills and partly via the kidneys. Freshwater fish were chosen for these experiments in view of the extremely high levels of circulating calcitonin reported in the seawater adapted eels by Orimo et al. (1972), while the dose of hormone used (10 MRC U 100 g^{-1} body wt i.p.) was considered to approximate that found in the circulation of seawater eels by the same authors. Although calcitonin had no effect on ^{22}Na efflux, that of ^{45}Ca was significantly reduced by about 22% compared with control injected fish, leading to the conclusion that the hormone may at least have covert effects on Ca turnover in this species.

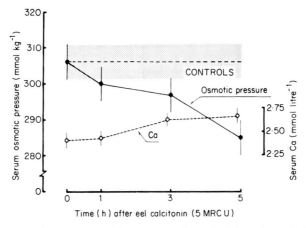

FIG. 29. Decrease in serum osmotic pressure after administration of eel calcitonin in *A. japonica*. Each point indicates the mean of determinations in 9 eels ($\pm$ s.e.). (From Orimo et al., 1972.)

Other workers have studied the effects of mammalian calcitonin in killifish (*F. kansae*) (Fleming et al., 1973). In these experiments it was reported that a single injection of 0·3 MRC mU g^{-1} body wt of porcine calcitonin stimulated ^{45}Ca uptake and altered the distribution of isotope between calcified (bone, skin and scales) and soft, tissue compartments in freshwater adapted fish (see Table XVI). The stimulation of Ca uptake was only apparent during summer months and disappeared in the winter when the Ca uptake of control fish was only about a third that of summer fish.

It appeared that control fish in winter distributed a greater proportion of the isotope into bone compared with that in summer fish while the converse was found with the proportion of isotope in skin and scales. In both groups of fish a similar proportion of the isotope appeared in soft tissues. Nevertheless, caution should be exercised when attempting to compare data in this way from separate groups of fish. Somewhat surprisingly in both groups of fish, a smaller proportion of the isotope appeared to enter the hard tissues following calcitonin treatment, than remained in the soft tissues. The bone response became refractory in winter but the response of skin and scales was greater than in summer treated fish.

It is difficult to reconcile the hard tissue responses reported in these fish with the known responses of mammalian bone to calcitonin and it is desirable that these experiments should be repeated with pure calcitonin preparations of fish origin and with regard to the alternative hard tissue component i.e. phosphate, in which case, the data on mineral uptake could become very significant (see Chapter 14).

TABLE XVI

Effect of porcine calcitonin on the Ca metabolism of male killifish (*F. kansae*)

	Summer-treated fish		Winter-treated fish	
	CT	C	CT	C
Net uptake μgCa g^{-1} dry wt h^{-1}	31·2 ± 1·8	18·3 ± 1·5	5·44 ± 0·68	5·53 ± 0·61
% ^{45}Ca distribution				
Bone	41·1 ± 0·9[a]	45·9 ± 1·3	38·1 ± 1·9	39·9 ± 2·2
Skin and scales	24·7 ± 2·1[a]	29·4 ± 1·2	26·3 ± 1·3[a]	36·6 ± 3·1
Soft tissues	34·2 ± 2·1[a]	24·7 ± 2·1	35·6 ± 1·9[a]	23·5 ± 2·3

Data are ± s.e. CT = calcitonin injected (0·3 MRC mUg^{-1} body wt), C = control injected.
[a] Significant difference $p < 0.02$ between CT and C injected groups.
(From Fleming et al., 1973.)

Different results have been reported in trout (*S. gairdnerii*) maintained in deionized water when the bones became demineralized by about 25% and hypocalcaemia developed after a time. Treatment with porcine calcitonin (2 MRC mU g^{-1} body wt 2 d^{-1}) for 3 weeks reversed the demineralization but had no effect on plasma Ca levels (Lopez et al., 1971). In a later paper, Lopez et al. (1976) investigated the action of salmon calcitonin on bone histology in immature eels. Chronic dosage (3 MRC mU g^{-1} body wt three times a week for 7 weeks) increased osteoblastic apposition and mineralization of the intercellular matrix in vertebral bone, but osteoclastic resorption and osteocytic osteolysis appeared to be unaffected by the hormone. A converse treatment, ultimobranchialectomy, completely stopped osteoblastic activity and mineralization, but again, did not appear to influence osteoclastic or osteocytic activity. Injections of carp pituitary extract, which caused massive hypercalcaemia and bone demineralization, were completely blocked by simultaneous calcitonin treatment, suggesting an antagonistic role of the pituitary and ultimobranchial on bone mineralization.

Pang et al. (1971a) extracted ultimobranchial tissues from two species of lungfish (*Neoceratodus forsteri* and *Lepidosiren paradoxa*) and found that these were hypocalcaemic when tested in young rats. There is only one report, however, of calcitonin having been tested in the Dipnoi or Choanichthyes (lungfish and coelacanths), that of Urist et al. (1972) who failed to demonstrate a hypocalcaemic response to the hormone in these species.

There are no reports of calcitonin being tested in the two lesser orders of Actinopterrygii (Holostei and Chondrostei). The Chondrostei (sturgeons) would be of particular interest in view of the highly degenerate nature of their bone.

ULTIMOBRANCHIAL EXTIRPATION

All experiments involving ultimobranchialectomy in the class Osteichthyes reported so far have been restricted to the order Teleostei, but the efficacy of this procedure is still far from clear. Thus in Japanese eels (*Anguilla japonica*) it was reported that destruction of the glands by electrocautery resulted in a small, but significant, increase in plasma Ca level, while plasma phosphate was unaffected (Chan, 1969). In a subsequent paper, the same author published completely contradictory data i.e. a significant fall in plasma phosphate and a small decrease in plasma Ca (Chan, 1972) with no explanation of the discrepancy in these results. However, Lopez et al. (1976) also obtained significant increases of around 25% in plasma Ca level within 2 weeks of ultimobranchialectomy in silver eels maintained in Ca enriched tap water, the levels declining to normal by 4 weeks. According to Chan (1972), ultimobranchialectomy resulted in a transient (1 week) decrease in urine phosphate, while urine Ca was elevated and remained so for 2 weeks after the surgery. In the same paper he reported atrophic, histological changes taking place in the corpuscles of Stannius following the surgical procedure. Some of the cells contain aldehyde fuchsin positive granules which are normally released into the lumen of the corpuscles. Two weeks after surgery the granules accumulated around the periphery of the cells but were not released. By the fourth week the cells showed marked regression, they clumped together and appeared to be loaded with granules. By the sixth week the cells were fully regressed and contained no granules. It is not clear whether this represents a removal of a trophic action of calcitonin on the corpuscles of Stannius, or reflects a change in plasma Ca level which, in turn, affected the corpuscles. Lopez et al. (1976) were unable to observe any such atrophic changes in the Stannius corpuscles following ultimobranchialectomy in eels.

Interpretation of results obtained following ultimobranchial "extirpation" in teleost fish are difficult since it is not always certain that the procedure has resulted in either complete abolition of calcitonin synthesis and secretion or, alternatively, a transient increase in release of this hormone. In the eel (*A. japonica*) the ultimobranchials consist of simple paired or unpaired tubular structures located immediately posterior to the ventral septum according to Chan (1972) who carried out exhaustive histological checks *post mortem* to ensure that the glands had been totally destroyed. There is evidence, however, from the studies of Orimo et al. (1972) that in the same species, the calcitonin secreting C cells are diffusely distributed within the oesophagus and pericardium (see Fig. 30). The data depicted in this figure indicate that it would be extremely difficult, if not impossible, to remove more than a fraction of the calcitonin secreting cells following "ultimobranchialectomy". In my experience

106 CALCIUM REGULATION IN SUB-MAMMALIAN VERTEBRATES

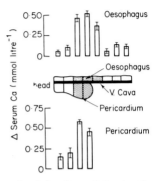

FIG. 30. The distribution of calcitonin activity in the ultimobranchial body of *A. japonica*. The oesophagus and pericardium were cut into small pieces (5 mm thickness) and the hypocalcaemic activity in each piece was compared with the rat bioassay. Each column indicates the mean of 5 assay rats ($\pm$ s.e.). (From Orimo et al., 1972).

(unpublished data) working with *A. anguilla*, it was concluded that complete electrocautery of the ultimobranchials could not be achieved without causing a fatal haemorrhage. Further studies are clearly needed to ascertain the effectiveness of ultimobranchialectomy in eels, and, since plasma calcitonin levels are easy to assay in these species, it should not be difficult to measure the levels of circulating hormone before and after surgery. Until the efficacy of the surgical technique has been demonstrated, it will be impossible to interpret the results of studies using this method.

Fenwick (1975) carried out partial ultimobranchialectomy in goldfish (*Carassius auratus*). When these fish were subsequently acutely transferred from fresh to 30% sea water, plasma Ca levels increased more than those of sham-operated controls (see Table XVII). Furthermore, plasma chloride (but not sodium) levels were also increased. These latter results corroborate those of Orimo et al. (1972) involving calcitonin injection into eels (see above), indicating a possible role for calcitonin in osmoregulation.

PLASMA CALCITONIN LEVELS

Plasma calcitonin levels in fish were first reported by Dacke et al. (1971). In species such as goldfish (*C. auratus*), these are so high (see Fig. 21) as to be detectable by bioassay. Similarly, Deftos et al. (1972), using radio-immunoassay techniques, found high levels of the hormone in species such as salmon (*Oncorhynchus nerka*), in which they may be as high as 66 mU ml^{-1} (see Chapter 5).

In their studies of calcitonin function in Japanese eels (*Anguilla japonica*), Orimo et al. (1972) used a sensitive bioassay procedure to measure the calcitonin content of both ultimobranchial tissue and plasma during transfer of the fish from a fresh- to seawater environment (see Figs 31 and 32).

TABLE XVII

Effect of ultimobranchialectomy on some blood parameters in freshwater adapted goldfish 20 h after acute transfer to 30% sea water[a]

Operation	Treatment Environment	Number	Sodium	Plasma conc. (mmol litre^{-1}) Calcium	Chloride	Hematocrit
Intact	Fresh water	10	158 ± 3.4	3.7 ± 0.3	118 ± 8.4	36 ± 2
	30% Sea water	10	172 ± 2.8	4.0 ± 0.3	137 ± 7.5	38 ± 3
Sham-operated	Fresh water	14	163 ± 2.6	3.5 ± 0.4	123 ± 4.7	35 ± 3
	30% Sea water	12	175 ± 2.2	4.1 ± 0.2	142 ± 5.7	41 ± 2
UBX[b]	Fresh water	6	160 ± 3.1	3.5 ± 0.2	128 ± 6.2	17 ± 4[c]
	30% Sea water	8	171 ± 2.5	5.4 ± 0.5[d]	168 ± 7.9[d]	19 ± 6[c]

[a] Values mean ± s.e.
[b] Partially ultimobranchialectomized.
[c] Significantly lower than values from control groups in either medium: $p < 0.01$.
[d] Significantly greater than values from control group in the same medium: $p < 0.05$.
(From Fenwick, 1975.)

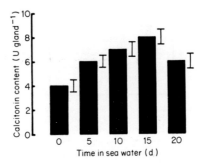

FIG. 31. Increase in calcitonin content of ultimobranchial of *A. japonica* transferred from fresh water to sea water. Each column indicates the mean of 3 assays (rat hypocalcaemic bioassay). (± s.e.) (From Orimo et al., 1972.)

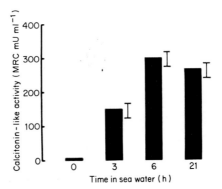

FIG. 32. Increase in plasma calcitonin-like activity of *A. japonica* transferred from fresh water to sea water. Plasma was pooled from 3 eels respectively and hypocalcaemic activity measured by rat bioassay. Bracket indicates 95% confidential limit of each essay. (From Orimo et al. 1972.)

The ultimobranchial calcitonin content of these fish increased about twofold over a 15 day period and then showed a slight decline. Plasma calcitonin levels were measured over a period of 21 h following transfer to sea water and a fifteenfold increase in these levels was apparent within 3 h, compared with the already extremely high calcitonin levels (10 MRC mU ml^{-1}) in the plasma of freshwater eels. The stimulus for this increase in calcitonin synthesis and secretion is probably the higher level of environmental Ca in sea water compared with fresh water.

Other evidence for the role of environmental Ca in regulating ultimobranchial function in Osteichthyan fish derives from studies of ultimobranchial histology in coral grazing species such as the parrotfish (*Sparisoma cretense*). The histological picture in these glands, as well as that of the Stannius corpuscles, suggests a high level of functional activity which is correlated with a high intake of dietary Ca.

It will be of interest in the future for studies on the effects of other seawater constituents such as magnesium, phosphate or total osmolality on plasma calcitonin levels, to be carried out in fish.

Changes in plasma calcitonin and Ca levels of migrating salmon have been reported by Watts *et al.* (1975). Their results, which are summarized in Fig. 33, demonstrated that both plasma Ca and calcitonin levels are consistently higher in female than male salmon. Plasma Ca levels showed a steady decline throughout the migratory transition from sea water to fresh water and through the

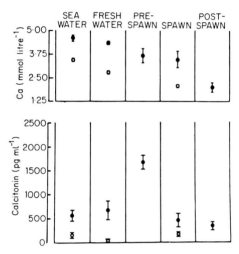

FIG. 33. Changes in plasma calcitonin and Ca during migration of salmon. (○) represents determinations in male salmon and (●) represents determinations in female salmon (± s.e.). (From Watts *et al.*, 1975.)

various phases of reproduction. Female fish, however, exhibited a pre-spawning surge in plasma calcitonin levels which fell back as spawning commenced. Unfortunately, Watts *et al.* did not obtain data for pre-spawning plasma calcitonin levels in male salmon. No clear correlation between plasma calcitonin and Ca levels was found in this study, suggesting that some other factor is responsible for the pre-spawning surge of plasma calcitonin in female fish. Apparently, goldfish (particularly females) show hyperplastic changes in ultimobranchial activity during the reproductive season (Oguri, 1973). Furthermore, plasma calcitonin levels in migratory salmon can be correlated with changes in the growth and resorption of bone occurring at this time (Watts *et al.*, 1975; Lopez and Deville, 1971). The pre-spawning surge in fish plasma calcitonin levels is similar to that found in female birds (see Chapter 13). While injections of oestrogen do not increase plasma calcitonin levels in salmon, according to Copp (1976), there are no data concerning possible effects of androgens, which, in birds, are shown to markedly influence the circulating hormone level (see Chapter 13). The role of gonadal hormones in regulating calcitonin synthesis and secretion in vertebrates has been discussed in more detail in Chapter 5.

There are still many questions concerning the role of calcitonin in the Osteichthyes. While this hormone appears to have some effects on the Ca metabolism, these are often covert. Calcitonin also has intriguing effects on osmoregulation suggesting some role during the migration of euryhaline species. It is possible, however, that these merely reflect secondary responses to changes in Ca metabolism at osmoregulatory membrane surfaces. Thus Bornancin et al. (1972) have shown that monovalent ion exchanges across the gill are particularly susceptible to changes in the Ca ion concentration at this site.

Some of the most interesting responses to calcitonin in the Osteichthyes concern the calcified tissues i.e. bone and scales. It is unfortunate that to date, no attempt has been made to study turnover of the phosphate ion in such tissues in fish. Part of the discussion in Chapter 14 is concerned with the possibility that bone and calcitonin evolved in fish in response to the need for phosphorus conservation.

There seems little doubt that calcitonin has some metabolic function in bony fish, but whether or not this is concerned primarily with Ca regulation remains to be seen.

Hypocalcin and the Stannius Corpuscles

The corpuscles of Stannius are present in most bony fishes where they have well-known effects on osmoregulation, probably through a mechanism analogous to the juxtaglomerular apparatus in kidneys of higher vertebrates (Bentley, 1976). It has also been recognized since 1964, that these tissues play a role in regulation of plasma Ca levels in bony fish, possibly through production of a hypocalcaemic factor "hypocalcin" (see Chapter 5).

STANNIUS CORPUSCLE EXTIRPATION

As with the ultimobranchial gland and calcitonin, most of the research concerning the role of the Stannius corpuscles in plasma Ca regulation in fish has been done with the teleosts, despite the fact that these organs reach their greatest development in the holostean order.

Fontaine (1964) was the first to demonstrate that removal of the Stannius corpuscles in the eel (*A. anguilla*) results in a rise in the plasma Ca level of one- to fourfold (see Fig. 34). This species is particularly useful for such studies since the two corpuscles are discrete and easily accessible for surgery, in a position just dorsal to the urinary bladder.

Similar responses were observed in goldfish (*C. auratus*) (Ogawa, 1968), Japanese eels (*A. japonica*) (Chan, 1969) and American eels (*A. rostrata*)

(Butler, 1969). The hypercalcaemia associated with Stanniectomy, appears within one week of surgery, but disappears after about two months (Fontaine, 1974). Plasma phosphate levels generally fall slightly following the operation (Fontaine, 1967), but plasma magnesium levels show no change after this procedure (Pang, 1971). Histological studies on the vertebrae of eels (*A. anguilla*) after Stanniectomy, revealed marked changes compared with those of normal fish, in particular, osteoclastic activity being greatly reduced

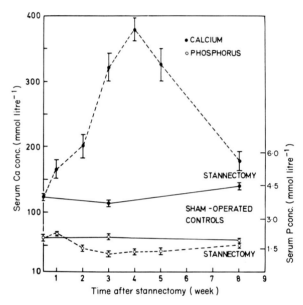

FIG. 34. The effect of ablation of the corpuscles of Stannius upon the serum Ca and phosphorus concentrations in the eel. (From Simmons, 1971; data after Fontaine, 1967.)

(Lopez, 1970a, b). Fontaine *et al.* (1972) reported that just when the hypercalcaemic response following Stanniectomy in eels is most marked, there is a significant increase in the Ca concentration of the bones. In eels, the surgical procedure is also reported to result in hypertrophy of the ultimobranchial gland, an effect which can probably be attributed to the stimulatory effect of hypercalcaemia on this organ (Chan, 1972). Some workers have reported that in eels, Stanniectomy causes a decreased urinary Ca content (Rankin *et al.*, 1967; Chan, 1972). Results from other workers, however, are at variance with these findings, in that a raised urinary Ca excretion is found (Butler, 1969; Fenwick, 1974). Fenwick carried out a very detailed series of studies using the American eel (*A. rostrata*). He measured net Ca excretion in these fish two weeks after Stanniectomy when maintained in environments of differing Ca concentration. Net Ca excretion was increased compared with sham-operated

TABLE XVIII

Urine Ca in stanniectomized eels after 14 days in water with different Ca concentrations[a]

Group and parameters	Environmental Ca con. (mmol litre^{-1})			
	7·82	0·45	0·02	0·0012
Sham-operated controls				
Urine flow (ml kg^{-1} day^{-1})	24·4 ± 3·1	26·5 ± 2·7	23·1 ± 5·3	52·6 ± 3·8[b]
Ca (mmol litre^{-1})	1·8 ± 0·2	1·6 ± 0·4	1·7 ± 0·4	1·9 ± 0·2
Ca excretion (mg kg^{-1} day^{-1})	1·7 ± 0·02	1·7 ± 0·04	1·6 ± 0·09	4·0 ± 0·03[b]
Stanniectomized				
Urine flow (ml kg^{-1} day^{-1})	22·4 ± 2·8	24·4 ± 3·4	26·4 ± 3·2	58·4 ± 4·8[b]
Ca (mmol litre^{-1})	3·0 ± 0·4[c]	2·7 ± 0·4[c]	2·8 ± 0·2[c]	1·8 ± 0·4
Ca excretion (mg kg^{-1} day^{-1})	2·6 ± 0·04[c]	2·6 ± 0·05[c]	2·9 ± 0·03[c]	4·2 ± 0·08[b]

[a] Values are mean ± s.e.; N = 10 in all cases.
[b] $p < 0.01$ compared to the same group in all other Ca environments.
[c] $p < 0.01$ compared to control groups in the same Ca environment.
(From Fenwick, 1974.)

controls when maintained in all but the most acalcic of environments. In the latter medium, both net urine flow and Ca excretion were markedly increased when compared with those of fish in environments with a higher Ca concentration (see Table XVIII). Certainly the hypercalcaemic response to Stanniectomy could not be accounted for by an increased retention of Ca from tubular urine. Furthermore, the urinary Ca loss was not increased significantly over controls for those fish maintained in acalcic water and neither did the fish show a hypercalcaemic response in this medium. This indicates that removal of the Stannius corpuscles does not result in mobilization of internal Ca reservoirs such as bone, a finding consistent with that of Stanniectomy inhibiting osteoclastic activity (Lopez, 1970a, b). On the basis of these results, Fenwick (1974) suggested that the corpuscles of Stannius probably exert their calcaemic effects by altering the balance of Ca exchange between fish and their external environment. Since the fish in these experiments were not fed and did not drink significant quantities of their media when held in fresh water, he concluded that the effects of the corpuscles of Stannius on Ca exchange are most likely mediated at the branchial level, so that following Stanniectomy, there is an overall increase in Ca influx through the gills. The role of gills as target organs in Ca exchange has been discussed in Chapter 6.

REPLACEMENT THERAPY

Fontaine (1964) was the first to demonstrate that injections of saline extracts of Stannius corpuscle into eels (*A. anguilla*) could be effective in reversing the hypercalcaemia produced by the surgical procedure. Similar results were obtained by Fenwick and Forster (1972) who retransplanted the corpuscles into previously Stanniectomized eels.

Recently, preliminary attempts to extract and purify the hypocalcaemic factor, or factors, from teleost Stannius corpuscles have been made by Pang *et al.* (1974). This group, using bioassay based on the hypocalcaemic responses of killifish (*F. heteroclitus*), has successfully identified a proteinaceous substance (hypocalcin) in killifish (*F. heteroclitus*) and cod (*Gadus morhua*) corpuscles, which has hypocalcaemic effects separate from those of any osmoregulatory substances in the corpuscles. The assay used by Pang *et al.* is rather insensitive and recently, Ma and Copp (1978) have developed a more sensitive assay based on inhibition of a Ca^{2+}/Mg^{2+} ATPase, isolated from the gill plasma membranes of rainbow trout. This assay has enabled extraction and purification of an active protein of mol. wt approx. 4000.

The experiments of Ogawa (1968) are at some variance with those of Pang and his colleagues, since Ogawa finds that treatment of Stanniectomized goldfish (*C. auratus*) with angiotensin II reverses both the hyponatraemia and hypercalcaemia caused by the operation. These results would suggest that the hypocalcaemic function of the Stannius corpuscles is not necessarily separate from that of sodium regulation and osmoregulation, although perhaps the effect of angiotensin II is analogous to that of hypocalcin.

It will probably be some time before assays of sufficient sensitivity are developed to enable the estimation of circulating hypocalcin levels in fish. Meanwhile, studies on the histological and biological properties of Stannius corpuscles have given an indication of the physiological role of this organ. It is well known that changes in the histological characteristics of the Stannius corpuscles occur in fish transferred to environments of differing salinity and that these changes are indicative of a role for these glands in osmoregulation and sodium balance (Bentley, 1976; Pang, 1973). Similarly, histological changes in the glands may be induced by immersion of fish in environments of similar osmotic constitution, but differing Ca concentration. Thus the activity of Stannius corpuscles in killifish (*F. heteroclitus*) appears greater in normal sea water than in Ca deficient sea water. Electron microscope studies of the corpuscles in the normal seawater adapted fish revealed a high degree of synthetic and secretory activity as indicated by proliferation of rough endoplasmic reticulum, hyperactive Golgi apparatus, granular depletion and presence of lysosome-like bodies. This evidence of hyperactivity was not present in glands from fish maintained in Ca deficient sea water for four or more weeks (Pang, 1973).

Vitamin D_3 Metabolites

Vitamin D_3 is undoubtedly present in the Osteichthyes and from the work of Henry and Norman (1975), we may conclude that in this class, the metabolic modification of the vitamin to its active metabolite or metabolites is similar to that in higher vertebrates (see Chapter 5). Unfortunately, the literature concerning the physiological role of vitamin D_3 and its metabolites in the Osteichthyes, or, for that matter, any other fish classes, is exceedingly scanty. This probably reflects the fact that administration of the vitamin, whether by oral or injection routes, apparently had no effect on plasma Ca levels (Urist, 1964, Moss, 1963). Poston (1969), however, fed massive doses of the vitamin (370 000 i.u. 100 g^{-1} food) to fingerling trout (*Salmo trutta*) maintained in fresh water, and was able to produce a hypercalcaemic response.

Apparently, more vitamin D_3 is stored in the livers of marine teleosts than in freshwater species. Thus values ranging from 25 i.u. g^{-1} liver in herring (*Clupea harengus*) to 250 000 i.u. g^{-1} liver in tuna (*Thunnus orientalis*), both marine species, have been reported, while the levels in freshwater species, are negligible (Urist, 1976b). It appears that part, at least, of this vitamin complement derives from non-dietary sources in species such as the cod (*G. morhua*) and must, therefore, be synthesized by the fish from a precursor 7-dehydrocholesterol (Blondin et al., 1967). It seems, that bony fish of the order Teleostei possess a metabolic pathway for vitamin D_3 metabolism similar to that of higher vertebrates which results in the production of a metabolite 1,25-dehydroxycholecalciferol. In higher vertebrates, this substance is considered to be essentially hormonal in character and is released in response to calcaemic or trophic stimuli (see Chapter 5). In mammals, and probably other classes, 1,25-dihydroxycholecalciferol affects gut, bone and, to a lesser extent, kidney, resulting in hypercalcaemia. We can only speculate at present on the role of this substance in osteichthyan Ca metabolism.

Recently, MacIntyre et al. (1976) demonstrated that i.p. injection of 1,25 dihydroxycholecalciferol (6n mol 100 g^{-1}) into eels (*A. anguilla*) results in an increase in the plasma phosphate level of about 10%. The 24,25 metabolite had no such effect. This finding is of particular interest in view of the possible evolutionary significance of phosphorus regulation in vertebrates (see Chapter 14).

In higher vertebrates, the gut appears to be the predominant target organ for 1,25-dihydroxycholecalciferol. In fish, the gut is of less importance in the uptake of Ca from the environment than it is in higher vertebrates. Berg (1968) studied the relative importance of food and environmental water as sources of Ca in the freshwater goldfish (*C. auratus*). Under natural conditions in which the Ca content of the freshwater medium is about 0·5 mmol litre^{-1} and dietary Ca is about 100 mmol kg^{-1}, it can be seen that only about 20% of Ca is derived directly from the food (Fig. 35). Even when the Ca

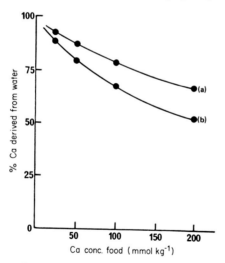

FIG. 35. Percentage of the total Ca exchange that is derived from the water at different concentrations of dietary Ca. Curve (a) is for natural lake water at Ca conc. ~ 0·5 mmol litre^{-1}. Curve (b) represents water at 0·1 mmol litre^{-1}. (From Simkiss, 1974; data after Berg, 1968.)

content of the water is as low as 0·1 mmol litre^{-1} and dietary Ca as high as 200 mmol kg^{-1}, it is clear that at least 50% of the Ca intake is derived from the water. Many workers consider that the diet contributes only between 2% and 10% of total Ca uptake under normal conditions (Simmons, 1971).

Since drinking is minimal in freshwater fish, it would appear that sites other than the gut are important for the uptake of Ca. These would be the gills and, to a lesser extent, skin and fin surfaces (see Chapter 6). In marine species the situation is rather different in that large quantities of Ca are present in the sea (see Table I) and bony fish living in this environment have to drink copious amounts of water in order to maintain an osmotic equilibrium (Maetz, 1974). It is not entirely clear how much of the total Ca uptake in these fish is via the gut (Simkiss, 1974), but in general, the contribution of gut in marine, as in freshwater fish, appears to be minimal. In seawater adapted *Tilapia mosambica*, for instance, blockage of the oesophagus has no effect on the rate at which these fish can absorb Ca, even though the control fish are able to swallow large quantities of the sea water (Reid *et al.*, 1959). Other studies involving Ca balance indicate that between 35% and 68% of swallowed Ca may be absorbed across the gut surface in marine species (Simkiss, 1974). Even these figures, however, are likely to be low when compared with the quantities of the ion absorbed across the gills and integument.

In bony fish it is unlikely that a role of 1,25-dihydroxycholecalciferol in increasing gut Ca transport would be of major importance in the regulation of Ca metabolism, although it will be interesting if in the future the metabolite is tested for effects in the gills and integument. In this respect, the relatively recent discovery of a specific binding protein for Ca in both gut and gills of teleost fish is encouraging (Chartier, 1973).

It will also be interesting to study the activity of the vitamin D_3 metabolites in fish bone and scales. The Chondrostei are an order of the class Osteichthyes in which bony tissues are almost completely degenerate. Members of this order, such as the sturgeon, have very low levels of vitamin D_3 in their livers according to Urist (1964). Furthermore, since neither the native vitamin nor the metabolic machinery necessary for its activation have been demonstrated in lower, non-bony classes of vertebrates, the question may be asked as to the principal role of the vitamin and its metabolites. Did it first evolve as a regulator of bone metabolism with its important effects on gut only developing at a later stage in evolution?

The Pituitary

The role of the pituitary gland in osteichthyan Ca regulation is still poorly defined, but its importance is becoming recognized. It seems that more than one of the pituitary hormones in fish can influence Ca regulation in both acute and chronic situations and it has been suggested that the gland may have a role which is similar to that of the parathyroid in tetrapods (see Chapter 5).

PITUITARY EXTIRPATION

In general, ablation of the teleost pituitary gland leads to a fall in the plasma Ca level (Simmons, 1971). Some workers find the hypocalcaemic response is most marked in fish which are adapted to media low in Ca content, whether these are fresh water (Fontaine, 1956) or artificial sea water from which the Ca has been omitted (Pang et al., 1971b). In the experiments of the latter group, male killifish (*F. heteroclitus*) were maintained in Ca deficient sea water for one month. Two groups of these fish were hypophysectomized while a third group was sham-operated. All groups were returned to the Ca deficient medium, but after four days one of the hypophysectomized groups was returned to normal sea water. The results (Table XIX) indicate that while plasma Ca and phosphate levels were appreciably lowered in hypophysectomized fish maintained

TABLE XIX

Effect of hypophysectomy on Ca and phosphate metabolism of killifish (*F. heteroclitus*) maintained in normal or Ca deficient sea water

Group	Plasma Ca (mmol litre^{-1} ± s.e.)	Plasma Pi (mmol litre^{-1} ± s.e.)
SW-Ca sham Hypex	2·29 ± 0·10	3·66 ± 0·35
SW-Ca Hypex	1·34 ± 0·08[a]	2·49 ± 0·20[a]
SW-Ca hypex Returned to SW	3·04 ± 0·29	2·84 ± 0·14
SW sham Hypex	3·68 ± 0·13	2·93 ± 0·22
SW Hypex	3·22 ± 0·24	2·93 ± 0·32

SW is sea water adapted, SW-Ca is Cal deficient sea water adapted, Hypex is hypophysectomized.
[a] Significantly different from control group ($p < 0.05$).
(From Pang et al., 1971b.)

in Ca deficient sea water, they were more or less corrected when the fish were returned to normal sea water. Most of the hypophysectomized group also exhibited symptoms of tetany, but subsequently recovered. Those groups maintained in Ca replete sea water showed little response to hypophysectomy.

In a more chronic situation, removal of the pituitary gland in fish can result in stunting of growth and demineralization of the otoliths and scales (Simmons, 1971). In some experiments the hypophysectomized fish developed renal lithiasis, the time course of which was linked to the period of defective calcification of the scales.

More recently, Brehe and Fleming (1976) studied the effects of hypophysectomy on ^{45}Ca turnover in the acellular boned killifish (*F. kansae*). Hypophysectomy reduced the rate of Ca turnover in terms of both ^{45}Ca uptake and efflux in fish during the summer months; it had no effect, however, on the (already) low turnover levels of winter fish. Since the experiments were carried out in male fish, the results could not be attributed to changes in Ca turnover associated with ovulation. Furthermore, hypophysectomy in summer fish caused a doubling of the half-times for influx of radiocalcium into the Ca reservoirs of bone, skin and scales, and soft tissues, whereas it had no such effect in winter fish. Similarly, the operation had significant effects on the distribution of both radiocalcium and total Ca within hard and soft tissues (see Table XX). Thus in summer fish, following hypophysectomy, less label was taken up by acellular bone and more remained in the soft tissue compartment.

TABLE XX
Effect of hypophysectomy on the distribution of total Ca and radiocalcium in male killifish (*F. kansae*)

Radiocalcium	Hypophysectomized	Controls
May and June		
Bone	47·0 ± 0·7[a]	50·6 ± 0·7
Skin/scales	34·1 ∓ 1·0	35·0 ± 0·9
Soft tissue	18·9 ± 0·9[a]	14·4 ± 1·0
December		
Bone	49·6 ± 0·7[a]	43·3 ± 0·6
Skin/scales	31·3 ± 0·6	32·9 ± 0·9
Soft tissue	19·1 ± 0·6[a]	23·8 ± 1·1
Total Ca		
May and June		
Bone	77·5 ± 0·8[a]	80·4 ± 0·9
Skin/scales	19·9 ± 0·7[a]	17·3 ± 0·9
Soft tissue	2·6 ± 0·2	2·4 ± 0·1
December		
Bone	81·8 ± 0·5	81·2 ± 0·5
Skin/scales	15·1 ± 0·6	16·4 ± 0·6
Soft tissue	3·1 ± 0·3	2·4 ± 0·9

Data presented as mean % distribution of ^{45}Ca and ^{40}Ca ± s.e.
[a] Significantly different from controls ($p < 0.05$).
(From Brehe and Fleming, 1976.)

In the winter fish, however, more label was taken up by the acellular bone in hypophysectomized fish, while less remained in the soft tissues. The quantity of label in the skin and scales did not appear to be affected by hypophysectomy in any season. These results were reflected to some extent by total Ca content of the tissues, but in summer fish, the skin and scales showed an increase in total Ca content following hypophysectomy, while paradoxically, in winter fish, neither bone nor skin and scales showed any response in total Ca content following the surgery.

Interpretation of these results, while difficult, is not impossible. It is likely that the pituitary hormone, or hormones, act at least at two different target organ levels. These could include mucosal epithelial tissues involved in direct Ca exchange between body fluids and environment such as gills, gut, renal tract and also the calcified tissue reservoirs such as bone and scales. Furthermore, probably more than one pituitary hormone is involved in Ca regulation in these fish. Thus prolactin might be directly involved in Ca exchanges at mucosal surfaces as well as regulating vitamin D_3 metabolism (see Chapter 5), which, in turn, might regulate Ca exchange both at mucosal and calcified tissue levels. Other pituitary hormones could include corticotrophin, which, as seen in Chapter 5, has some structural homology with parathyroid hormone as well as the recently identified parathyroid-like hypercalcin (see Chapter 5). The data from Brehe and Fleming (1976) are, therefore, extremely difficult to interpret and further work is needed to clarify the situation. This should obviously include studies with purified pituitary hormones (preferably of teleost origin) and involve selected target tissues *in vitro* or isolated preparations *in situ*.

Replacement Therapy

Early experiments concerning pituitary replacement therapy in teleost fish have centred on the role of growth hormone. Hence purified bovine growth hormone or fish pituitary (which contains some thyrotrophic hormone) can restore growth, and scale and otolith mineralization in brown trout and killifish (Simmons, 1971). More recent work has focused on the role of pituitary hormones such as prolactin and corticotrophin. The results with prolactin are equivocal; in the eel (*A. anguilla*) Olivereau and Chartier-Baraduc (1966) found that prolactin therapy to some extent corrected the hypocalcaemia following hypophysectomy if the fish were maintained in normal fresh water. In fish maintained in distilled water, prolactin was ineffective in reversing the hypocalcaemia; indeed, these fish showed an even greater hypocalcaemia compared with control fish by 10 days (see Fig. 36).

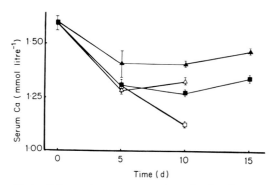

Fig. 36. The effect of hypophysectomy and prolactin therapy on plasma Ca concentrations in the eel (*A. anguilla*). (■) hypophysectomized—fresh water, (△) hypophysectomized—distilled water, (▲) hypophysectomized + prolactin—fresh water, (□) hypophysectomized + prolactin—distilled water. (Data from Olivereau and Chartier-Baraduc, 1966.)

Killifish maintained in a Ca free, artificial sea water and fed Ca deficient food, either prolactin, corticotrophin, or, to some extent cortisol, could correct the hypocalcaemia resulting from hypophysectomy. Melanophore stimulating hormone was, however, ineffective (see Table XXI). Prolactin was also effective

TABLE XXI
Effect of pituitary replacement therapy on serum Ca levels of hypophysectomized killifish (*F. heteroclitus*)

Groups		Serum Ca (mmol litre^{-1} ± s.e.)
(1) Sham controls saline		2·36 ± 0·15[a]
(2) Hypex + saline		1·75 ± 0·13
(3) Hypex + cortisol	2·5 µg g^{-1}	2·18 ± 0·06[a]
(4) Hypex + corticotrophin	0·05µg g^{-1}	2·50 ± 0·27[a]
(5) Hypex + prolactin	5 µg g^{-1}	2·49 ± 0·14[a]
(6) Hypex + melanophore stimulating hormone	2·5 µg g^{-1}	1·97 ± 0·08

Hypex is hypophysectomized.
[a] Significantly different from group 2 ($p < 0.05$).
(From Pang, 1973.)

in reversing the hypocalcaemia, even at 10% of the above dose, although corticotrophin was not (Pang, 1973).

In freshwater fish, secretion of prolactin is considered to predominate and help in survival of the fish, while cortisol secretion is considered to be of more importance in seawater fish. Thus the level of prolactin stored in the pituitary of the teleost *Poecilia latipinna* is six times greater in fish living in fresh water than those in sea water, while the activity of the pituitary eta cells indicates that more of the hormone is being released in the freshwater fish (Ball and Ingleton, 1973). Plasma cortisol levels in eels introduced from fresh water to sea water increase by about 40% within two days, but after a few more days they decline to levels which are similar to those of freshwater fish (Forrest *et al.*, 1973).

As for the site of calcaemic actions of corticotrophin, cortisol and prolactin, we can, at present, only speculate. These hormones also have major actions on sodium and water regulation in fish. The general effect of cortisol appears to be to increase sodium efflux in seawater fish, while prolactin prevents sodium loss and increases water excretion in freshwater fish. The major site for the sodium regulating effects of these hormones is the gill, where they affect ATPase activity and membrane permeability; the gut and kidney may also be involved to a lesser extent (Bentley, 1976) and it seems feasible that the hypercalcaemic effects could also be modulated by the gill. Whether or not the calcaemic effects represent part of a more general electrolyte regulation by this organ, or whether a specific and separate effect on Ca transfer across the branchial mucosa exists, remains to be seen.

If immature eels are treated with carp pituitary extract they develop hypercalcaemia accompanied by intense osteoclastic and perilacunar bone resorption (Lopez et al., 1968). We are reminded here of the role of a bone response in parathyroid mediated hypercalcaemia and of the structural homology of parathyroid hormone and corticotrophin. The recent discovery that pituitaries of fish (eel and cod) contain a peptide (hypercalcin) with a close immunological relationship with the amino-terminal (1–34) of mammalian parathyroid hormone (see Chapter 5) will stimulate renewed interest in the role of the pituitary in vertebrate Ca regulation. Is hypercalcin the major Ca regulating factor of the fish pituitary, at what stage in the vertebrate hierarchy did it evolve and which are its target organs?

Gonadal Hormones

Adult female osteichthyan fish, in common with female amphibians, reptiles and birds, exhibit seasonal variations in plasma Ca levels associated with their breeding cycles. The hypercalcaemia is correlated with maturation of the ovarian follicle and the influence of oestrogens which stimulate synthesis of yolk protein and, hence, an increase in the protein bound fraction of plasma Ca. Sex differences in plasma Ca levels in these fish were first reported by Hess et al. (1928) who found a range of 2·3–3·1 mmol litre^{-1} of Ca in male cod (*G. morhua*), while in mature females the range was 3·2–7·1 mmol litre^{-1}. The females with the highest plasma Ca content also had the most mature gonads. Similarly, in female killifish (*F. kansae*), plasma Ca levels increase threefold in the summer compared with winter levels (Fleming et al., 1964), this change not being found in male fish, although it can, however, be imitated in males by injecting oestradiol. Several other workers have demonstrated similar responses to oestrogens in other teleost fish (Pang, 1973) as well as in Dipnoi (Urist, 1976b).

Mugiya and Watabe (1977) demonstrated a resorptive effect of oestrogen on scales in killifish and goldfish. A single injection of oestradiol benzoate (5 μg g^{-1} body wt) resulted within 2 days in a 25% decrease in ^{45}Ca content of scales in killifish (*F. heteroclitus*) which had been labelled with isotope 6 days previously. The response lasted for at least 14 days. A slightly slower response, but of even greater amplitude, was obtained in goldfish (*C. auratus*) and similar but smaller responses were elicited from the bone (vertebrae and jawbone) in these fish. While these results are at variance with the anabolic effect of oestrogen on avian and mammalian bone, they do seem to relate not only to the hypercalcaemic action of oestrogens in fish, but also to changes in calcified tissues found in some fish prior to spawning (see above). It is possible that these responses reflect an action of oestrogen not directly on bone, but on the secretion of calcitrophic hormones such as calcitonin (see above and Chapter 5).

There is little evidence to indicate any direct effect of androgens on plasma

Ca levels (Pang, 1963), although Woodhead (1968) has reported a seasonal increase in this parameter in male cod (*G. morhua*) in sea water. Male salmon (*Salmo solar*) spawning in fresh water are reported to exhibit decreased plasma Ca levels (Fontaine *et al.*, 1969). It was suggested by Pang (1973) that these findings could reflect an inability of the male fish to regulate plasma Ca at the time of spawning and that since sea water is high in Ca, while fresh water is low in this ion, the plasma Ca values merely correspond to differences in the environment. It is possible that these changes also represent variations in the activity of hormones such as calcitonin and hypocalcin which have been discussed previously.

Conclusion

It is quite clear that the Ca levels of osteichthyan extracellular fluids are regulated to a finer degree than those of more primitive vertebrates. We are presented with a bewildering range of hormones in this fish class which seem to play some role in Ca regulation. Thus hormones from the pituitary, ultimobranchials, Stannius corpuscles and gonads are all shown to have some effect on osteichthyan Ca metabolism. Often these responses appear covert and it is worth asking whether some of the primary effects of these hormones might be concerned with other electrolytes such as phosphate or magnesium, both of which pose major problems to fish and which apparently have no specific regulatory systems. Surprisingly, the vitamin D_3 metabolites, which play one of the most important Ca regulating roles in higher vertebrates, have little effect in fish, although the studies on this substance to date have been extremely limited in these species.

Can one identify an endocrine system which is specifically involved with regulating the Ca ion in bony fish? The answer, at present, is no, yet there is no shortage of possible candidates for this role.

11. Calcium Regulation in Amphibia

The class Amphibia occupies an important position in vertebrate phylogeny and its representatives show the first major habitation of a terrestrial, as opposed to an aquatic, environment. This transition from water to land must have presented many metabolic problems to these animals, not least of which would be the loss of a supply of environmental calcium (Ca) which could be absorbed directly across the major exchange sites of integument and gills. This continuous supply of environmental Ca is partly replaced by a more intermittent dietary supply of the ion and is probably reflected by an increasingly important role of the gut as an organ for Ca absorption (see below). Calcium uptake via the skin in this class may also achieve greater priority than in fish, perhaps in part replacing the now defunct gills. A second problem associated with colonization of land and, hence, the development of airbreathing, is that of acid-base balance. Fish are able to regulate their blood pH by using the gills as organs for excretion of ions such as ammonium and bicarbonate in exchange for sodium and chloride (Maetz, 1974). Loss of the gill in adult amphibia complicates acid-base regulation, particularly in the order Anura, and this has led to development of a major exchangeable alkali reserve, i.e. the $CaCO_3$ deposits of the endolymphatic sacs.

Colonization of land by the Amphibia is incomplete; the adult is amphibious, while the larval forms are aquatic. Metamorphosis of larval to adult forms has many interesting repercussions, not least concerning the Ca metabolism which undergoes a "crisis" at this time, according to Simkiss (1967).

It is in the class Amphibia that parathyroid glands make their first phylogenic appearance. In most of the tetrapods these glands appear to be essential for survival so that their surgical removal results in tetanic convulsions and ultimately in death. Apart from the first appearance of parathyroid glands in amphibia, other major endocrine changes occur, including the disappearance of Stannius corpuscles (assuming they were present in ancestral osteichthyan forms) and the first recognizable role for vitamin D_3 in vertebrate Ca metabolism.

Plasma Calcium Levels

Plasma Ca concentrations in amphibians are similar to those of higher vertebrate classes (Simkiss, 1967). There is evidence, however, that plasma Ca is

124 CALCIUM REGULATION IN SUB-MAMMALIAN VERTEBRATES

less well regulated in amphibia, since both male and female adults of this class show marked seasonal fluctuation in their plasma Ca levels. In the bullfrog (*Rana catasbeiana*), these have been documented by Studitsky (1945) quoting his colleague Kovalsky. Spring plasma Ca levels in males are 2·10 mmol litre^{-1} falling to 1·35 mmol litre^{-1} in the summer and rising to 1·85 mmol litre^{-1} in winter. These changes are accompanied by fluctuations in the activity of the parathyroid glands (see Chapter 5). Similar and more recent data obtained

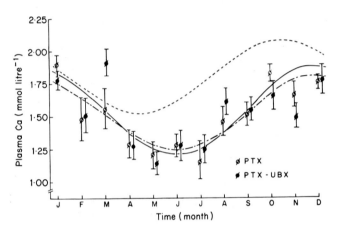

FIG. 37. Annual plasma Ca concentration as a response to parathyroidectomy and parathyroidectomy-ultimobranchialectomy (5–8 days post-operative). The periodic function for PTX frogs (N = 105) is expressed as: $Ca_p = 6·158 + 1·346 \cos(30t - 360)$; $r = 0·791$; $p = 0·001$. For PTX-UBX frogs (N = 81), the function is: $Ca_p = 6·087 + 1·120 \cos(30t - 360)$; $r = 0·665$; $p = 0·005$. Both groups were not significantly different from one another, but the response as expressed by the plasma Ca was significantly ($p < 0·001$) different from the control curve (– – – –). Each point is mean ± s.d. (N = 5–10 frogs). (From Robertson, 1977.)

from *Rana pipiens* by Robertson (1977) are depicted in Fig. 37. In other species, the seasonal plasma Ca fluctuations are less marked: in the clawed toad (*Xenopus laevis*), male Ca levels fluctuate between 2·08 and 2·45 mmol litre^{-1} (Swarenstein and Shapiro, 1933).

Plasma and urine phosphate levels in *R. pipiens* also vary seasonally and are inverse to each other; plasma phosphate is high in summer and low in winter while urine phosphate is low in summer and high in winter (McWhinnie and Lehrer, 1972).

Parathyroid Hormone

PARATHYROID GLAND EXTIRPATION IN ANURA

It is generally recognized that the parathyroids first appear in amphibia. The results of experiments involving parathyroid extirpation in this class are equivocal. Studitsky (1945) suggested that parathyroidectomy in frogs has no deleterious effect in so far as the animals did not show symptoms of tetany and eventual death. Other workers have suggested the operation in frogs causes a lowering of the plasma Ca concentration, although this is not necessarily accompanied by tetany (Waggener, 1930; Cortelyou, 1960; Cortelyou and McWhinnie, 1967). Other effects of parathyroidectomy in frogs include rises in plasma phosphate and urinary phosphate excretion (Cortelyou, 1960, 1962a). The latter effect is the opposite of that found in mammals and is, therefore, somewhat surprising, since the hyperphosphataemic response to parathyroidectomy in mammals is usually accredited to increased renal phosphate resorption (Kenny and Dacke, 1975).

Recently it was shown by Robertson (1974) that parathyroidectomy in frogs not only reduces urinary Ca loss to a small extent, but also that of sodium and water; this occurred irrespective of whether the frogs were kept in fresh water or transferred to either deionized water or 0·05 M sodium chloride. These responses appeared to be independent of glomerular filtration rate which was unchanged by the operation, indicating a response at the tubular level.

PARATHYROID INJECTION IN ANURA

Injections of 10 U.S.P. units of bovine parathyroid hormone into frogs resulted in an increased plasma Ca concentration and a fall in urinary Ca within two hours, but urinary phosphate was unaffected (Cortelyou, 1962b). Since the plasma and urine phosphate responses probably reflect increased bone mobilization, it would seem that the amphibian kidney does not exhibit the major response to the hormone which is seen in higher tetrapods.

In general, parathyroid hormone injection into amphibia (at least Anurans), increases bone resorption as it does in mammals; Schlumberger and Burk (1953) reported that under the influence of the hormone, resorption of bone in *R. pipiens* was expressed as demineralization of the cortex of the long bones and calvariae. Simultaneously, the X-ray density of the $CaCO_3$ deposits in the lime sacs surrounding spinal ganglia was increased, leading these investigators to suggest that the lime sacs act as a Ca reservoir for the maintenance of blood Ca levels during periods of bone resorption.

LAVAGE STUDIES IN ANURA

Evidence for a role of parathyroid hormone in regulating bone mineral turnover in the bullfrog has also been provided by Yoshida and Talmage (1962). The frogs were either warm-acclimatized at 22 °C or cold-acclimatized at 4 °C and a Ca free peritoneal lavage was initiated 24 h after injection of ^{45}Ca in order to stimulate the parathyroids. In this situation about twice as much ^{45}Ca was mobilized from the Ca reservoirs of warm-acclimatized than from cold-acclimatized frogs. Parathyroidectomy of the warm-acclimatized frogs depressed ^{45}Ca mobilization to the same level as that in cold frogs. The bone histology picture was also followed in these experiments; while bones of normal frogs yielded a relative osteoclast count of 10 cells per field ($\times$ 100), the value was increased by eightfold in those animals which had been stressed with the Ca free lavage. Cold frogs and parathyroidectomized-warm or cold frogs had identical osteoclast counts, which were about 70% less than those in stimulated warm-acclimatized animals. These data indicate that either removal of glands, or exposure to cold, bring the frogs to a metabolic minimum, i.e. parathyroidectomy depresses bone metabolism of warm-acclimatized frogs to a reduced rate characteristic of cold exposure.

PARATHYROIDS IN URODELES

In urodelan amphibia it is more difficult to induce parathyroid hormone responses such as osteoclastic and osteocytic bone resorption (Bélanger and Drouin, 1966, Simmons et al., 1971). In the newt (*Cynops pyrrhogaster*) parathyroidectomy caused a marked fall in plasma Ca levels within five days, but this procedure was ineffective in another urodelan, *Hynobius nigrescens* (see Table XXIII, p. 136). Replacement therapy with parathyroid tissue from either species was effective in raising plasma Ca levels in *C. pyrrhogaster* but not in *H. nigrescens*. Plasma sodium and magnesium levels were unchanged by these procedures (Oguro and Uchiyama, 1975).

SEASONAL CYCLES

It has been recognized since the early work of Waggener (1929) that the anuran parathyroid gland undergoes a seasonal cycle of cytolysis and regeneration. In bullfrogs (*R. catasbeinea*), the breakdown begins in February and regeneration is complete by May. This appears to be closely related to the seasonal fluctuations in plasma Ca levels which are found in both males and females of this species (see above). The detailed histological changes occurring in the cyclical degeneration of anuran parathyroids have been described by Cortelyou and McWhinnie (1967). These degenerative changes, which occur in winter in several species of frogs and toads, include nuclear chromatocytolysis and pycnosis, vacuolation and loss of structural integrity. Some species' differences exist, however, with respect to the morphological detail and timing of the seasonal changes. While the precise relationship between plasma Ca levels and seasonal changes in parathyroid activity is not clear, it would seem to be a direct one, i.e. the plasma Ca level is rising at a time when the parathyroid gland is active. This was demonstrated particularly well in a recent paper by Robertson (1977). He studied the effects of parathyroidectomy and ultimobranchialectomy, either separately or in combination, on the seasonal cycle of plasma Ca levels in the frog (*R. pipiens*). Data from this paper are shown in Fig. 37. Control frogs exhibited a sinusoidal pattern in their plasma Ca levels which were at a low point around April–May and high around October–November. Ultimobranchialectomy did not affect these levels (at least within 5–8 days of surgery), but parathyroidectomy or parathyroidectomy in combination with ultimobranchialectomy was effective in lowering them (within 5–8 days) during the period May until November, i.e. during the ascending phase in the seasonal cycle of plasma Ca. In contrast, parathyroidectomy or parathyroidectomy-ultimobranchialectomy appeared to have little effect on the plasma Ca levels during the period from December–April, which suggests that the ascending phase of the Ca cycle is associated with increasing parathyroid gland activity, while the descending phase is related to a shutdown in activity of this gland.

Detailed studies of the bone metabolic changes elicited by parathyroid hormone under varying conditions of season and temperature were made by McWhinnie and Cortelyou (1967). During summer or periods of warm acclimatization, hormone treatment increased the rate of carbohydrate metabolism in bone (see Table XXII), while bone citric acid levels which could be involved in bone solubilization (see Chapter 6) were elevated (see Fig. 38). Hence in winter, oxidation of the substrate was inhibited while similar responses followed parathyroidectomy. Replacement therapy with parathyroid

TABLE XXII
Influence of parathyroid hormone on substrate utilization by femur homogenates of *R. pipiens*

Substrates	Δ O.D. 340 mμ mg^{-1} protein per 3 min (NADP $\rightarrow$ NADPH)				
	Normal		PTX		
	Summer	Winter	Summer	PTX + PTH	% Increase
			Diaphyses		
Glucose-6-phosphate	1·33	0·31	0·35	0·53	52%
6-Phosphogluconate	0·51	0·23	0·17	0·23	32%
			Epiphyses		
Glucose-6-phosphate	1·39	0·53	0·43	0·58	34%
6-Phosphogluconate	0·52	0·15	0·13	0·18	41%

(From McWhinnie and Cortelyou, 1967.)

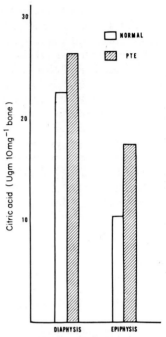

FIG. 38. The effect of mammalian parathyroid extract (PTE) on the content of citric acid in the femur of *R. pipiens* during the summer. Frogs were killed 24 h after treatment with 20 U.S.P. units of PTE. Both diaphysis and epiphysis show significant ($p < 0.05$) increases in citric acid content after PTE treatment but that of the epiphysis is proportionately much greater than that of the diaphysis. (From McWhinnie and Cortelyou, 1967.)

hormone was partially successful in restoring the reduced values in parathyroidectomized frogs (see Table XXII). At the same time, pre-existing sulphated mucopolysaccharides of "old" bone matrix were degraded, while epiphyseal zones enlarged and exhibited accelerated or abnormal elaboration of sulphated components of osteoid or cartilage matrix (Fig. 39).

While unfortunately lacking in statistical evaluation, these latter results are consistent with those found in mammalian bone where parathyroid hormone stimulates synthesis of mucopolysaccharide and subsequent incorporation of S^{35} into bone (Bernstein and Handler, 1958).

Skin Response

Parathyroid hormone influences Ca translocation by frog skin by causing significant increases in influx and decreases in efflux of the ion across this tissue *in vitro* (see Chapter 6). These responses presumably occur independ-

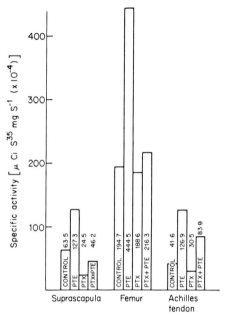

FIG. 39. The effect of parathyroidectomy (PTX) and parathyroid extract (PTE) upon the levels of sulphur-35 in skeletal and collagenous tissues of *R. pipiens*. (From McWhinnie and Cortelyou, 1967.)

ently of the vitamin D_3 system, unlike the gut response to parathyroid hormone in vertebrates (see Chapter 6). Such a response in amphibians aids the conservation of Ca or associated ions in these species.

It appears, then, that the physiological effects of parathyroid hormone in anuran amphibians, at least with respect to plasma Ca and bone response, are similar to those of higher tetrapods, including mammals. A major difference may lie in the marked seasonal variations in parathyroid activity, with ambient temperature seemingly acting as the modulating influence. The role of this hormone in amphibian acid-base balance is considered below.

Calcitonin and the Ultimobranchials

Amphibian ultimobranchial glands are well developed and may even be active in the larval (tadpole) stages of species such as bullfrogs (*R. catasbeiana*), according to Robertson (1971). Morphologically they consist of an irregular follicular structure with a central ductless cavity lined with pseudostratified epithelium. Cytological evidence indicating an endocrine function for this gland in amphibia was reviewed by Robertson (1965), before the discovery that it represents the seat of calcitonin secretion. Like the amphibian parathyroid, the ultimobranchial gland in this class exhibits a seasonal cycle of activity and quiescence. In glands of winter frogs, endocrine secretory activity cannot be demonstrated (Robertson and Bell, 1965), nor can basal stem cells be readily detected (Robertson, 1967).

ULTIMOBRANCHIAL EXTIRPATION

Removal of ultimobranchial glands in frogs (*R. pipiens*) results initially in a hypercalcaemic response (see Fig. 40) but after about eight weeks plasma Ca levels return to normal and may even fall below control values (Robertson, 1968b, 1969a, b). Robertson considers the latter response to result from excessive mobilization of bone and lime sac Ca, to the extent that these organs become Ca depleted. Increased osteoclastic activity becomes apparent in bone following the operation (Robertson, 1969a), while the urinary Ca content is also elevated (Robertson, 1969b).

A special feature of ultimobranchial glands in amphibians and probably other classes, is the presence of unmyelinated and myelinated axons within the pericapillary space, with the unmyelinated axons terminating upon the basal regions of mature secretory cells. The vesicular elements comprising these bulbous terminations identify these axons as sympathetic in nature (Robertson, 1968b). He carried out experiments in male *R. pipiens* in which the ultimobranchial glands were effectively denervated by removal and subsequent

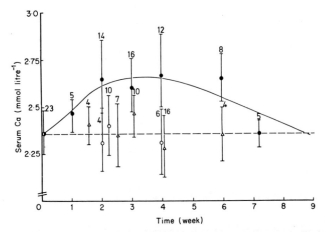

FIG. 40. The effects of total ultimobranchialectomy on frog serum Ca levels over a period of 8 weeks, is depicted by the curve drawn to the solid circles. Normal values ● and (- - -). Bilateral homeotopic transplants (△) and heterotopic transplants (○) indicate Ca levels in animals in which viable glands were recovered. All values represented ± s.d. with the number of animals indicated in each group. (From Robertson, 1968.)

autoplastic transplantation at a site remote from their original position. In the denervated glands, ergastoplasm and Golgi membranes exhibited hypertrophy. Under hypercalcaemic conditions, however, the autotransplants were able effectively to maintain plasma Ca levels (see Fig. 40), indicating that the process of secretion by these tissues is primarily an intrinsic cellular activity independent of the innervation. These results have led Robertson (1968) to suggest that the sympathetic axons may be inhibitory, thus permitting a depression of glandular function during periods of hypocalcaemia. Nevertheless, the concrete evidence that the ultimobranchial innervation has an inhibitory function has yet to be produced.

In a previous section concerning the seasonal plasma Ca cycle in frogs, it was suggested that, while either parathyroidectomy or parathyroidectomy-ultimobranchialectomy resulted in a lowering of plasma Ca levels during a period from April–November, ultimobranchialectomy alone had no apparent effect, which may seem paradoxical since ultimobranchial extirpation should result in a transient hypercalcaemia (see above). This latter effect, however, only occurs about 2–6 weeks after surgery, and in the experiments concerned with seasonal effects of ultimobranchialectomy Robertson (1977) carried out plasma Ca determinations 5–8 days after surgery. Perhaps this was too short a time to demonstrate any effect of calcitonin depletion on the seasonal cycle of plasma Ca so that an influence of this hormone on that cycle cannot be ruled out. Since the amphibian ultimobranchial gland shows a similar histological pattern of activity and quiescence to the parathyroid, there would not appear to be any clear link between ultimobranchial function and plasma Ca levels such as there is with the parathyroids.

Robertson (1974) has investigated the role of the ultimobranchial as well as parathyroid glands on sodium and water excretion in frogs (*R. pipiens*). Ultimobranchialectomized frogs maintained in fresh water displayed a transitory diuresis within 3 days of surgery. When the frogs were submitted to a mild environmental saline load, the urinary excretion rates were increased by about 75% in controls, but only by 15% in the ultimobranchialectomized group. Removal of the ultimobranchial glands in frogs can be viewed as producing a hyperparathyroid situation since the increase in urinary excretion rate over the 2 day post-operative period was transitory with a return to normal by 7 days. In view of effects of the parathyroids on sodium and water excretion reported in the same paper (Robertson, 1974), the transitory response in ultimobranchialectomized frogs probably reflects the effect of unopposed parathyroid secretion which declines as plasma Ca levels increase. Under a mild sodium load, total urine sodium loss was no greater than in controls, while in deionized water, the increase in urinary sodium loss after ultimobranchialectomy could be accounted for by an increased glomerular filtration rate. In all three aqueous media, ultimobranchialectomized frogs consistently exhibited hypercalcuria independently of the total sodium loss. This led Robertson (1974) to the conclusion that while the parathyroid effects on renal Ca loss are related to renal tubular reabsorption of sodium, the ultimobranchial can affect renal Ca excretion without influencing sodium, independently of the parathyroids.

Robertson (1969c) also produced evidence for an ultimobranchial influence on amphibian bone. He studied the healing processes of closed fractures of tibiofibulae in *R. pipiens* by X-ray and histological techniques following ultimobranchialectomy. The repair processes of fractured bones in the first 4 weeks revealed similar patterns in both intact controls and ultimobranchialectomized frogs. There was extensive cartilage formation around the fracture with some ossification at the periphery of the cartilage. After 8 weeks the fracture sites of control frogs exhibited more extensive invasion and development of endochondral ossification than did ultimobranchialectomized frogs. Osteoclasts were also obvious at the distal segment of the fracture, particularly in controls. After 12 weeks periosteal bone was well developed in controls but only poorly represented in ultimobranchialectomized frogs. It seems that ultimobranchialectomy and, therefore, calcitonin deprivation has a long-term rather than acute effect on bone, at least with regard to its histological picture. Cellular activity is greatly reduced with a concomitant reduction in calcification between 8 and 12 weeks after surgery. Since this period is associated with increased osteoclastic activity, Robertson suggested that the repair processes may be retarded by high serum Ca levels which, in turn, may influence parathyroid hormone production so that calcitonin does not necessarily have any direct influence on bone in amphibians. This hypothesis seems unlikely, however, on two points. Firstly, the hypercalcaemic response following ultimobranchialectomy has corrected itself in the frog by 7 or 8 weeks after surgery (see Fig. 40) and probably reflects a decrease in parathyroid activity at this time which would also tend to reduce bone cellular activity. Secondly, calcitonin has well-established effects on mammalian bone and also those of other vertebrate species, which are consistent with the responses found in amphibians (see above). In mammals, for instance, calcitonin inhibits the appearance of multinucleate osteoclasts (Reynolds *et al.*, 1968).

CALCITONIN INJECTION

Few studies have been carried out on the effects of injected purified calcitonin in amphibians. There is one report (Boschwitz and Bern, 1971) that a single injection of porcine calcitonin caused a significant lowering of plasma Ca levels within 30 min in marine toads (*Bufo marinus*), although chronic administration of the hormone (0·05 MRC U per day for 37 days) caused significant hypercalcaemia.

FIG. 41. (a) X-ray of control frog kept in fresh water for 2 weeks demonstrating scant deposition of Ca in the paravertebral lime sacs ($\rightarrow$). (b) Frog injected with 100 000 i.u. vitamin D_2 and maintained in 75 mmol litre^{-1} $CaCl_2$ for 7 days. Note the accumulation of Ca as paired, segmental densities along the vertebral column ($\rightarrow$). (c) Frog injected with 200 000 i.u. vitamin D_2 and maintained in 75 mmol litre^{-1} $CaCl_2$ for 14 days. Note extensive filling of lime sacs ($\rightarrow$). (d) Frog injected as in (c) above; followed by 3 weeks in fresh water. Lime sacs are extensively filled. Note the decreased density of membrane bones in skull and at epiphyseal plates of the long bones ($\rightarrow$). (From Robertson, 1968.)

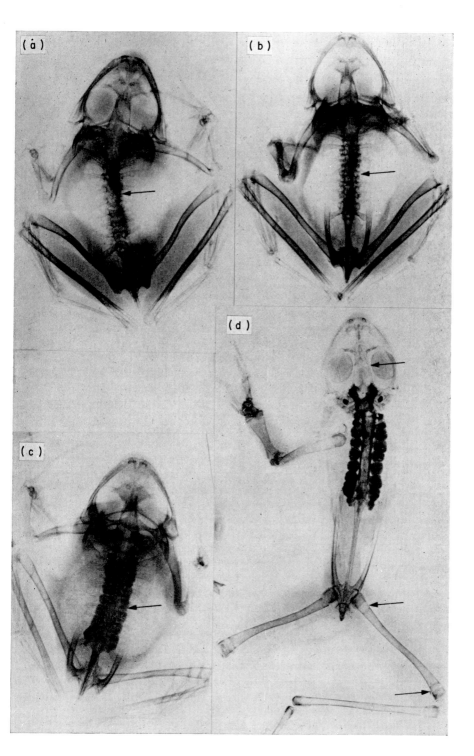

Vitamin D_3 Metabolites

A little more is known of the physiological role of vitamin D_3 in the class Amphibia than in fish, but nevertheless there is a relative paucity of information. Similarly, little is known about the absorption of Ca or phosphate across frog gut, although there have been studies on Ca transport in frog gastric mucosa (Forte and Nauss, 1966).

GENERAL EFFECTS

Earlier, Schlumberger and Burk (1953) demonstrated that following treatment with synthetic vitamin D_2 of frogs maintained in water to which Ca salts had been added, Ca was deposited at storage sites (bone and endolymphatic sacs) and renal calcification occurred. They interpreted these results as a demonstration that vitamin D is effective in enhancing the intestinal absorption of Ca. Similar findings by Robertson (1968) have substantiated this report, by treating frogs (*R. pipiens*) for 2 weeks with a total intramuscular dose of 200 000 i.u. vitamin D_2; the plasma Ca level rose from a normal level of 2·3 mmol litre^{-1} to 3·8 mmol litre^{-1}. During these experiments the frogs were maintained in 0·8% $CaCl_2$ solution and the paravertebral lime sacs were found to have undergone extensive filling under the hypercalcaemic conditions (see Fig. 41).

EFFECTS ON GUT

More recently, (Robertson, 1975a) studied the effect of vitamin D_3 on *in vitro* Ca transport by frog everted gut sacs. The gut sacs were separated by ligature into duodenal and jejunal-ileal segments. Frogs were dosed intramuscularly with 25 000 i.u. vitamin D_3 and serosal/mucosal Ca ratios measured in excised guts at various times after this treatment. Calcium transport was increased maximally 2 days after vitamin D_3 dosage. Furthermore, transport of the ion was greater in the duodenal than the jejunal-ileal portion of gut (see Fig. 42). An additional observation was that sensitivity of the gut to vitamin D_3 is greater in September, when plasma Ca levels are high, than in March, when they are lower (see Table XXIII).

Robertson (1975b) has also studied the effect of vitamin D_3 on intestinal Ca transport *in vitro* in ultimobranchialectomized and parathyroidectomized frogs; a dose of 100 000 i.u. of the vitamin was used in this study and again the guts were divided into jejunal-ileal and duodenal segments. The results (Fig. 43) show that the vitamin increased Ca transport in both segments of gut, although the increase was transient (2 days) in the jejunal-ileal segment compared with 4 days in the duodenal segment. In the jejunal-ileal segment Ca

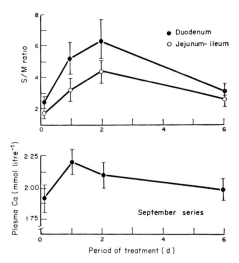

FIG. 42. Response of frogs to 25 000 i.u. vitamin D_3 and Ca ions during September (± s.d.). The serosal/mucosal (S/M) ratio of duodenal segment was significantly higher ($p = 0.05$) than jejunal-ileal segment on the first and second day which was concomitant with a maximal hypercalcaemia after the first day (4 animals in each group). (From Robertson, 1975a.)

transport was reduced by parathyroidectomy or ultimobranchialectomy-parathyroidectomy at 2 and 4 days. Parathyroidectomy alone also significantly decreased the 3 day response in the duodenal segment. Here Ca transport was enhanced following ultimobranchialectomy, at 4 days, compared with that in control operated frogs.

The results of this rather complicated series of experiments are of considerable interest. They indicate firstly that the two segments of the frog intestine can show differential responses to hormones in terms of Ca transport. In both segments vitamin D_3 enhances transport of the ion and the fact that parathyroidectomy inhibits this transport indicates a role of parathyroid hormone in frogs similar to that in mammals, i.e. facilitating the hydroxylation of 25-hydroxycholecalciferol to 1,25-dihydroxycholecalciferol in the kidney. The enhancement of Ca transport in ultimobranchialectomized frog intestinal segments could indicate an inhibitory effect on 25-hydroxycholecalciferol hydroxylation if it were found in both intestinal segments, but this response is only seen in the duodenal segment. The jejunal-ileal segments do not even show a tendency towards this response, suggesting that the effect of ultimobranchialectomy on duodenal Ca transport occurs independently of vitamin D_3. These results indicate that Ca transport in the frog intestine is mainly under the influence of vitamin D_3 with parathyroid hormone acting in a permissive role, allowing synthesis of the active metabolite of the vitamin. It is clear from the studies of Henry and Norman (1974), that the frog kidney, like

TABLE XXIII

Dose sensitivity of *R. pipiens* to vitamin D_3 in the presence or absence of Ca (0·05M $CaCl_2$) after 2 days of treatment during September–March

Time period	Dose vitamin D_3 (i.u. per frog)	Ca in water	Whole gut (S/M ratio ± s.e.)	Plasma Ca (mmol litre^{-1} ± s.e.)
September–October	0	—	3·1 ± 0·1	1·90 ± 0·05
	25 000	—	3·2 ± 0·1	1·85 ± 0·05
	25 000	+	5·0 ± 0·3[a]	2·05 ± 0·07[a]
December–January	0	—	2·0 ± 0·2	1·75 ± 0·05
	25 000	—	3·4 ± 0·2[a]	1·90 ± 0·07[a]
	50 000	+	3·2 ± 0·3[a]	1·88 ± 0·07[a]
March	0	—	2·4 ± 0·1	1·53 ± 0·05
	25 000	—	2·4 ± 0·1	1·70 ± 0·07
	50 000	+	3·2 ± 0·2[a]	1·80 ± 0·07[a]
	100 000	—	2·4 ± 0·2	1·70 ± 0·07
	100 000	+	4·4 ± 0·3[a]	2·00 ± 0·08[a]

[a] Significantly different from controls ($p < 0.01$).
(After Robertson, 1975a)

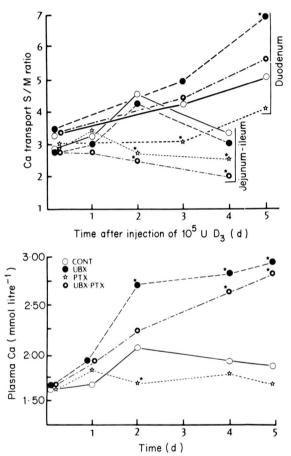

FIG. 43. Response of 4-day post-operative frogs to 2·50 mg (100 000 i.u.) vitamin D_3 and ingestion of 0·05M CaCl during month of March. Each point is mean of 4–6 frogs, (*) indicates $p < 0.01$ when compared to controls of duodenal or jejunal-ileal segments and controls of plasma Ca. Plasma Ca levels ($\pm$ s.e.) of frogs kept only in 0·05M CaCl after 3 days were as follows: Controls (1·75 $\pm$ 0·03); ultimobranchialectomized (UBX) (2·10 $\pm$ 0·05); parathyroidectomized (PTX) (1·51 $\pm$ 0·08) and UBX-PTX (1·90 $\pm$ 0·01). A significant ($p < 0.01$) hypercalcaemia with vitamin D^3 was seen only in UBX and UBX-PTX frogs. (From Robertson, 1975b.)

that of other vertebrate classes, possesses metabolic machinery necessary for the final hydroxylation step of vitamin D_3. Calcitonin appears to have an inhibitory effect on Ca transport which is quite specific for the duodenum and independent of vitamin D_3 metabolism.

LUNAR INFLUENCES

Not only is Ca transport in frog gut regulated by hormones, but this process also appears to be influenced by the phases of the moon. Robertson (1975c) has made a study of the diurnal and lunar periodicity of intestinal Ca transport and plasma Ca in the frog (*R. pipiens*). In all frogs, male and female, duodenal Ca transport as measured by the everted gut sac ratio method showed a marked diurnal fluctuation. Peaks of activity clearly occurred at dusk and dawn (see Fig. 44), when expressed as a monthly average for all frogs. If the

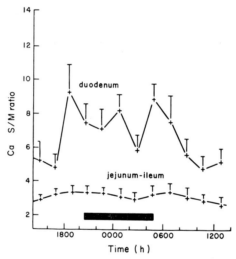

FIG. 44. Cumulative mean of intestinal Ca transport activity data for each 2h segment (N = 125) at the 4 lunar phases, to provide an "average" response during a lunar month. Maximal responses in the duodenum are at dusk and dawn with elevated activity during the nocturnal period. The jejunum-ileum exhibits relatively constant Ca transport activity throughout the 24 h period. Each point is mean ± s.e. (From Robertson, 1976.)

effects of lunar transit are superimposed onto the above graph a picture is seen as in Fig. 45. In this figure, the dotted lines indicate average diurnal Ca transport (Fig. 44) while solid lines indicate average Ca transport at particular phases of the moon and also at periods corresponding to high or low lunar transit. The shaded areas represent periods when the normal diurnal rate of gut Ca absorption is enhanced by a lunar transit and it is obvious that both high and low lunar transits significantly correlate with enhanced gut Ca transport. This effect was most marked if the moon was in its first or third quarter. Amphibia exhibit diurnal and lunar fluctuations in many aspects of their metabolism and it is likely that other classes of vertebrates may present similar rhythms (Robertson, 1976).

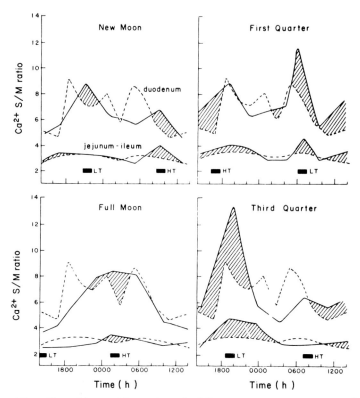

FIG. 45. The effect of lunar tides on intestinal Ca transport appears as an enhancement effect (hatched area) above the average monthly response (– – – – –). At each of the 4 lunar phases the corresponding lunar transit (HT is high transit, LT is low transit) has a high correlation ($r = 0.939$, $p = 0.001$) to each intestinal segment. (From Robertson, 1976.)

Pituitary Hormone

Reports of a possible pituitary involvement in amphibian Ca regulation date back to 1933, when Shapiro and Zwarenstein proposed that hypophysectomy could cause a decrease in serum Ca concentration, regardless of the gonadal condition, in the urodele *X. laevis*. More recently, Doughtery (1973) reported that pituitary implantation significantly raised the serum Ca level in the anuran *R. pipiens* in winter, when the serum Ca level is normally low compared with summer.

As well as studying the role of parathyroids in urodeles, Oguro and Uchiyama (1975) examined the influence of the pituitary on divalent ion metabolism in this order. Following hypophysectomy, plasma Ca levels in two species of newt, *C. pyrrhogaster* and *H. nigrescens*, were reduced by about 17% after 10 days, while plasma sodium and magnesium values were reduced even more: the results for *C. pyrrhogaster* are depicted in Table XXIV. This would appear

TABLE XXIV
Serum Ca, sodium and magnesium concentrations of *C. pyrrhogaster* after parathyroidectomy or hypophysectomy

Treatment		Ca (mmol litre^{-1})	Na (mmol litre^{-1})	Mg (mmol litre^{-1})
5 days	PTXS	2·80 ± 0·10	107·5 ± 1·5	1·17 ± 0·05
	PTX	1·50 ± 0·05^b	104·0 ± 3·0	1·29 ± 0·10
10 days	PTXS	2·92 ± 0·15	110·0 ± 1·0	0·86 ± 0·05
	PTX	2·12 ± 0·08^a	105·0 ± 3·2	1·17 ± 0·05^a
20 days	PTXS	2·52 ± 0·05	104·5 ± 2·0	0·96 ± 0·05
	PTX	2·20 ± 0·18^a	104·4 ± 0·5	1·00 ± 0·10
5 days	HYPXS	2·52 ± 0·13	106·0 ± 2·2	1·17 ± 0·05
	HYPX	2·61 ± 0·15	105·0 ± 1·3	1·21 ± 0·05
10 days	HYPXS	2·58 ± 0·05	105·2 ± 1·3	1·00 ± 0·05
	HYPX	2·26 ± 0·05^b	92·0 ± 1·6^b	0·79 ± 0·10^b
20 days	HYPXS	2·52 ± 0·08	103·7 ± 0·8	0·92 ± 0·05
	HYPX	2·25 ± 0·08^a	73·8 ± 3·0^b	0·67 ± 0·05^b

All data mean ± s.e.
PTX is parathyroidectomized, PTXS is sham-parathyroidectomized, HYPX is hypophysectomized. HYPXS is sham-hypophysectomized.
^{a} Significantly different from sham-operated controls ($p < 0.05$).
^{b} Significantly different from sham-operated controls ($p < 0.01$).
(After Oguro and Uchiyama, 1975.)

to be the only study of a pituitary influence on the plasma magnesium concentration in any class of sub-mammalian vertebrate and is, therefore, of considerable interest in ascertaining the regulation of this important ion (see Chapter 14).

As yet there are no reports of studies on the effects of administration of purified pituitary hormones such as prolactin, corticotrophin or the recently discovered hypercalcin, to amphibia. Such studies are clearly desirable at this time.

Other Hormones

There are no reports at present concerning the possible effects of fish Stannius corpuscle extracts on amphibian Ca regulation. Since amphibia do not possess Stannius corpuscles this is not surprising. It is to be hoped that fish hypocalcin will be tested for calcaemic effects in amphibia fairly soon although the results of such experiments would be mainly of theoretical interest.

The Endolymphatic Sacs and Acid-base Balance

CALCIUM CARBONATE RESERVES

Two hypotheses have been advanced concerning the function of enlarged paravertebral lime sac deposits in amphibia. The first is the "anatomical theory" suggesting a role protecting the spinal column, but this is poorly supported by experimental evidence. A second theory proposes that the large $CaCO_3$ deposits within these organs have a physiological function connected with the homeostatic processes in the blood (Simkiss, 1968). It has been suggested that these calcareous deposits can be withdrawn to assist in bone repair (Schlumberger and Burk, 1953) or to provide Ca for bone ossification during metamorphosis from larval to adult forms (Guardabassi, 1960; Pilkington and Simkiss, 1966), and there is good experimental evidence to support both these ideas. Another proposal is that the $CaCO_3$ deposits play a role in acid-base balance in amphibia, since exposure to high levels of environmental carbon dioxide leads to dissolution of the calcareous deposits of the lime sacs

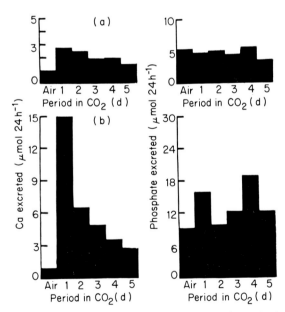

FIG. 46. Loss of Ca and phosphate ions from frogs in air, and when exposed to (a) 5% CO_2 + 95% O_2 (4 frogs) and (b) 10% CO_2 + 90% O_2 (4 frogs). Note the increase in Ca excretion which is not accompanied by increased phosphate excretion. (From Simkiss, 1968.)

(Sulze, 1942). Simkiss (1968) made a detailed investigation of the role of endolymphatic lime deposits in amphibian acid-base balance by exposing frogs (*R. temporaria*) to atmospheres containing high carbon dioxide levels (5% and 10% CO_2); under these conditions net Ca excretion was greatly increased while phosphate loss was minimal (see Fig. 46). He found that Ca loss under stimulus of respiratory acidosis was much larger than that found under hypocalcic (distilled water) conditions. Loss of phosphate, although rather variable, was not significantly greater under acidotic compared with hypocalcic conditions. This led Simkiss to the conclusion that the increased Ca loss in acidosis represents mobilization of $CaCO_3$ stores from the endolymphatic sacs. In the same study he made an estimate of blood bicarbonate and carbonate levels during respiratory acidosis. Bicarbonate levels were derived from the Henderson-Hasselbalch equation:

$$pH = pK_1 + \log \frac{HCO_3}{\alpha CO_2}$$

where pH and PCO_2 content of the frog blood are known and using values of 6·15 for pK_I at pH 7·66 and 25 °C, and 0·041 for α at 25 °C. From this he also derived blood carbonate values by using the logarithmic form of the law of mass action:

$$PCO_3 = PHCO_3 - pH + pK_2$$

and taking pK_2 as 10·22, which is the value for mammalian blood.

The changes in blood bicarbonate and carbonate levels following exposure to 5% CO_2 are shown in Fig. 47. Bicarbonate levels exhibited a persistent rise

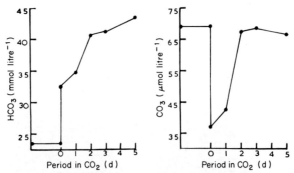

FIG. 47. Changes in bicarbonate and carbonate levels of the blood of frogs exposed to 5% CO_2 + 95% O_2 for up to 5 days. (From Simkiss, 1968.)

during respiratory acidosis while carbonate levels showed first a fall of around 50%, but then recovered within about 2 days. Simkiss (1968) has proposed the following equation to account for these changes:

CALCIUM REGULATION IN AMPHIBIA 143

$$CaCO_3 \text{ (solid)} \rightleftharpoons Ca^{2+} + CO_3^{2-}$$
$$\Updownarrow$$
$$HCO_3^- \rightleftharpoons H^+ + CO_3^{2-}$$

In this way, blood $CaCO_3$ can act as a buffer. The carbonate levels, which fall initially during acidosis, are quickly replenished by a mobilization of endolymphatic $CaCO_3$ deposits. This interpretation implies that these deposits are in equilibrium with the blood, a suggestion which is in keeping with the rich capillary network associated with the large endolymphatic sacs in frogs.

HORMONAL INFLUENCES

Robertson (1972) has investigated the effect of endocrine systems on gross exchange of the endolymphatic lime deposits. In the ultimobranchialectomized frog these structures become essentially depleted of $CaCO_3$ in two weeks, whereas in parathyroidectomized or parathyroidectomized-ultimobranchialectomized frogs they become filled. This suggests that parathyroid hormone is necessary for mobilization of the deposits while calcitonin has an inhibitory influence on their excessive mobilization. Similarly, the lime deposits increase during hypervitaminosis D_3 (Schlumburger and Burk, 1953), although this response might simply reflect a secondary response to the rise in blood Ca following treatment with the vitamin. The responses to ultimobranchialectomy and parathyroidectomy found by Robertson (1972) seem to represent a direct effect of calcitonin and parathyroid hormone on the endolymphatic $CaCO_3$ deposits since ultimobranchialectomy and parathyroid hormone cause, respectively, hypercalcaemia and hypocalcaemia in frogs (see above).

Exposure to high atmospheric carbon dioxide in intact, ultimobranchialectomized, parathyroidectomized or ultimobranchialectomized-parathyroidectomized frogs, resulted in marked changes in serum composition. Blood pH fell

TABLE XXV
Changes in acid-base states of blood in adult male frogs after 3 days of exposure to 10% CO_2 per 90% O_2 following a 3-day post-operative period

Group	pH	HCO_3^- (mmol litre^{-1})	PCO_2 (mm Hg)	$CO_3^=$ (μmol litre^{-1})	Ca (mmol litre^{-1})
Controls	7·70 ± 0·05	24 ± 3	15 ± 3	76 ± 10	1·75 ± 0·15
UBX	7·57 ± 0·05	27 ± 3	23 ± 4	63 ± 12	1·70 ± 0·13
PTX	7·38 ± 0·04	21 ± 3	30 ± 3	30 ± 6	1·15 ± 0·13
UBX-PTX	7·40 ± 0·01	21 ± 1	29 ± 3	34 ± 4	1·35 ± 0·13

UBX is ultimobranchialectomized, PTX is parathyroidectomized, UBX-PTX is ultimobranchialectomized-parathyroidectomized.
Values mean ± s.d.
(After Robertson, 1972.)

TABLE XXVI

Effect of glandular removal in adult male frogs during February–May on serum ions under "normal" environmental conditions

Group	pH	HCO$_3^-$ (mmol litre^{-1})	PCO$_2$ (mm Hg)	CO$_3^=$ (μmol litre^{-1})	Ca (mmol litre^{-1})
Normal (36)	7·73 ± 0·08	17 ± 2	11 ± 3	58 ± 6	2·33 ± 0·20
UBX-3 (14)	7·82 ± 0·08	19 ± 3	6 ± 4	76 ± 9	2·38 ± 0·18
UBX-21 (9)	7·75 ± 0·09	14 ± 3	8 ± 4	48 ± 10	1·65 ± 0·18
PTX-3 (11)	7·76 ± 0·09	15 ± 3	9 ± 4	55 ± 8	1·40 ± 0·13
PTX-21 (3)	7·76 ± 0·09	13 ± 3	7 ± 4	43 ± 8	1·33 ± 0·15
UBX-PTX-3 (3)	7·67 ± 0·08	15 ± 3	10 ± 4	44 ± 8	1·45 ± 0·13
UBX-PTX-21 (5)	7·70 ± 0·05	10 ± 2	5 ± 4	30 ± 10	1·64 ± 0·22

UBX is ultimobranchialectomized, PTX is parathyroidectomized, UBX-PTX is ultimobranchialectomized-parathyroidectomized.
Numbers after groups are post-operative days, number of animals in brackets.
Values mean ± s.d.
(After Robertson, 1972.)

abruptly in all groups within 6 h, but in controls this was corrected within 24 h. In ultimobranchialectomized, parathyroidectomized and ultimobranchialectomized-parathyroidectomized groups, compensation was slower, particularly in parathyroidectomized and ultimobranchialectomized-parathyroidectomized groups, which continued to exhibit an uncompensated respiratory acidosis and apparently raised PCO_2 after 72 h, at least when compared with the remarkably low control value for this parameter, as well as symptoms of tetany. The 72 h results are shown in Table XXV.

Robertson (1972) also investigated the relationship between calcitonin, parathyroid hormone and the endolymphatic sacs in acid-base balance of frogs under normal atmospheric conditions. The results indicate that both ultimobranchialectomy and parathyroidectomy gave rise to an alkalosis of respiratory origin as shown by the blood PCO_2 (Table XXVI). In ultimobranchialectomy-parathyroidectomy, a slight acidosis occurred at 3 days, at least when compared with the apparently, extremely alkalotic controls, with compensation occurring at 21 days, as shown by blood bicarbonate level and PCO_2.

Plasma carbonate levels were also derived in these experiments by the same method used by Simkiss (1968). These fell markedly in the parathyroidectomized as well as parathyroidectomized-ultimobranchialectomized animals during exposure to 10% atmospheric CO_2 (see Table XXV). This is consistent with a role for parathyroid hormone in mobilizing $CaCO_3$ from the lime sacs and with subsequent chemical dissociation of this substance along the lines suggested by Simkiss (1968) (see above). It is clear from these results that parathyroid hormone is necessary for mobilization of $CaCO_3$ buffer reserves from the lime sacs during respiratory acidosis. Robertson (1972) has proposed that this effect could be mediated locally in the lime sac epithelium by an increase in protons and bicarbonate ions, through the action of parathyroid hormone on carbonic anhydrase, thus effectively altering solubility of $CaCO_3$ to give a systemic increase in Ca and carbonate. In support of this concept he found that parathyroidectomy in *R. pipiens* results in a 60% depletion of lime sac carbonic anhydrase within 2 weeks. Ultimobranchialectomy had little effect on this enzyme. Calcitonin may, however, have some effect on lime sac $CaCO_3$ reserves since the transient rise in plasma Ca levels following ultimobranchialectomy, with a return to normal after several weeks (Fig. 40), seems to indicate mobilization of a finite Ca store, i.e. endolymphatic lime deposits. The inability of ultimobranchialectomized frogs completely to regulate pH during acute respiratory acidosis, as well as the apparent lack of this gland's effect on lime sac epithelial carbonic anhydrase activity, seem inconsistent with a role for calcitonin in inhibiting $CaCO_3$ mobilization. Such a role should result in marked but transient increases in blood buffer supply.

From the above discussion it is apparent that the endolymphatic lime deposits in anurans are intimately connected with acid-base balance. Perhaps this specialized evolutionary adaptation is a result of transition from an aquatic to terrestrial mode of life. Fish are able to regulate plasma pH by direct exchange of ions such as protons, ammonium and bicarbonate with sodium and chloride across the gills (Maetz, 1974). Consequently, they have no need for a large endogenous reserve of buffer base, and $CaCO_3$ stores in bone (or elsewhere) are lacking. Amphibia, at least of the order Anura, do not possess gills in the adult form, but instead have primitive lungs. They seem to have solved the acid-base problem by greatly extending the endolymphatic $CaCO_3$ buffer deposits. In higher tetrapods, this function may be taken over by the $CaCO_3$ deposits in bone (Blitz and Pellegrino, 1969), and probably also by development of lungs and kidneys, which become able to compensate for blood pH variations more efficiently than the comparable organs of amphibians.

A recent interesting study concerns the role of the toad (*Buffo marinus*) urinary bladder in acid-base regulation. This organ excretes protons and ammonium ions in response to acidosis. Frazier (1976) demonstrated that application of pure parathyroid hormone (10 μg ml^{-1}) to the serosal side of the *in vitro* toad bladder raised H$^+$ excretion by 50% above control values. There was no effect on NH$^+_4$ excretion and the increase in acid excretion appeared to be mediated by 3′,5′-cyclic AMP.

One pertinent question at this stage is whether parathyroid hormone evolved in this class in answer to a need for improved acid-base regulation rather than for Ca regulation. This will be considered in more detail in Chapter 14.

12. Calcium Regulation in Reptiles

Reptiles represent the mainstream of tetrapod evolution, being the descendants of vertebrates which first colonized land and, at the same time, having phyletic affinities with both mammals and birds. Many reptiles share with birds the problems associated with secreting large quantities of calcium (Ca) salts into calcareous eggshells surrounding their cleidoic eggs. Other reptiles lead a semi-aquatic life and face similar problems to amphibians. A few species live in the sea and must, therefore, cope with hypercalcic and hyperosmotic environments. Some reptiles also have well-developed dermal skeletons containing massive deposits of Ca. Unfortunately, though an interesting class for comparative physiologists to study, they have little economic or clinical relevance and there are few papers concerning reptilian Ca metabolism.

Blood Calcium Levels

Plasma Ca levels in several representative species of reptiles have been reported by Clark (1972); these range from 2·40 mmol litre^{-1} in a turtle, *Chrysemys picta*, to 3·42 mmol litre^{-1} in a snake, *Thamnophis sirtalis*, while a lizard, such as *Anolis carolinensis*, had a value of 2·61 mmol litre^{-1}. Plasma phosphate levels are also rather variable, ranging from 0·93 mmol litre^{-1} in *C. picta* to 1·95 mmol litre^{-1} in *A. carolinensis*. These values, from reptiles other than reproducing females, are similar to values in other tetrapods. Reproductively active females have higher plasma Ca levels (see Tables XIV and XXIX) by virtue of the yolk protein precursors circulating in the blood. There is no information currently in the literature concerning possible seasonal fluctuations in reptilian plasma Ca levels as there is for amphibians. Since reptiles (at least the extant species) are poikilotherms, one might expect such variations to occur in association with seasonal ebb and flow of metabolic activity.

Reptiles possess all the major hormones and tissues associated with Ca regulation in mammals. They may also possess endolymphatic deposits of CaCO$_3$, but these are generally less well developed than in the anuran amphibia.

TABLE XXVII
Effect of parathyroidectomy in reptiles

Group	Species	Normal serum Ca (mmol litre^{-1} ± s.e.)	Normal serum Pi (mmol litre^{-1} ± s.e.)	Time after parathyroidectomy (d)	% Change Ca	% Change Pi	Tetany?
Turtles	*Chrysemys picta* *Pseudemys scripta*	2·40 ± 0·06	0·93 ± 0·04	1–56	None	None	No
Lizards	*Varanus griseus*	3·08 ± 0·32	—	1–6	40↓	—	Yes
	Anolis carolinensis	2·61 ± 0·11	1·95 ± 0·13	3–4	56↓	72↑	Yes
Snakes	*Elaphe quadrivirgata*	3·25 ± 0·09	—	10	33↓	—	Yes
	Thamnophis sirtalis	3·42 ± 0·31	1·65 ± 0·12	10	60↓	44↑	Yes
				42	70↓	152↑	

(From Clark, 1972.)

Parathyroid Hormone

PARATHYROID EXTIRPATION

Removal of parathyroid glands in reptiles usually leads to a fall in serum Ca levels accompanied by tetany (see Table XXVII). Turtles, such as *C. picta* and *Pseudomys scripta*, appear to be more resistant to the surgery than other reptiles since the changes in serum Ca and phosphate levels are not apparent in this order. They do, however, show a marked fall in urinary phosphate (Clark, 1965). In the snake, *T. sirtalis*, parathyroidectomy results in a fall in serum Ca, the level of which remains low for several weeks after surgery, and tetanic convulsions are common in this species following surgery. Serum phosphate shows a corresponding rise after surgery (Clark, 1971). Similar results were found in the lizard (*A. carolinensis*), although serum Ca levels were very variable in this species (Clark *et al.*, 1969). The papers cited above also contain details of the anatomical and histological structure of reptilian parathyroids.

PARATHYROID HORMONE INJECTION

Administration of mammalian parathyroid extract generally has an opposite effect to that of parathyroidectomy in reptiles and large doses of the extract usually lead to increases in serum Ca, urine Ca and phosphate levels. In some cases a fall in serum phosphate is obvious (Table XXVIII). The time intervals used in these experiments were relatively long (1–7 days) and in the future it will be of interest to see if lower doses of parathyroid hormone are able to elicit the rapid, but transient, hypercalcaemic responses which are characteristic of amphibia and birds (see Chapters 11 and 13). Chronic treatment with mammalian parathyroid hormone (64–96 U.S.P. U) for 24 days increased the bone osteoclast count in the lizard *Lacerta agilis* according to Umansky and Kudokatzev (1951) and Clark (1965) obtained similar data in lizards and turtles. There are no reports concerning the effect of parathyroid hormone on other organs such as the gut or endolymphatic lime sacs as in amphibia. Since the parathyroids apparently have a profound effect on endolymphatic Ca deposits and, hence, on acid-base balance, it would be useful to study the effect of the hormone on these parameters in reptiles. Similarly, the substantial $CaCO_3$ fraction found in turtle skeletons (see Chapter 4) is worthy of further investigation.

TABLE XXVIII

Effect of parathyroid extract (PTE) in reptiles

Group	Species	Time after PTE administration	Total dose (i.u.)	% Change serum Ca	% Change serum Pi	% Change urine Ca	% Change urine Pi
Turtles	*Chrysemys picta* N, P	3 days	100–378	20↑NS	0.5↓NS	224↑NS	600↑S
Lizards	*Dipsosaurus dorsalis* N	7 days (1 injection per day)	105	20↑S	6↑NS	810↑S	1980↑S
	Sceloporus grammicus N	7 days (1 injection per day)	105	—	—	250↑S	280↑S
	Anolis carolinensis N	26 h	97	37↑NS	5↓NS	—	—
	Anolis carolinensis P	26 h (4 injections)	97	73↑S	36↓S	—	—
Snakes	*Thamnophis sirtalis* N	20 h	1.18 100 gbw⁻¹	17↑NS	46↓S	—	—
			2.36 100 gbw⁻¹	31↑S	17↓NS	—	—

N = Intact animals, P = Parathyroidectomized animals, NS = Not significantly different from control values, S = Significantly different from control values ($p < 0.05$), gbw = grams body weight.
(From Clark, 1972.)

Calcitonin

ULTIMOBRANCHIAL ANATOMY IN REPTILES

The ultimobranchial gland is characteristically present only on the left side of the neck, close to the caudal parathyroid gland; it tends to be small and diffuse and consists of scattered epithelial follicles in connective tissue (Clark, 1972). The function of this gland is not as clear as that of the parathyroids in reptiles; there are no reports in the literature of the effect of ultimobranchial removal in this class but this procedure would be difficult owing to the diffuse nature of the gland in these animals.

CALCITONIN INJECTION

Extracts of ultimobranchial tissue from reptiles have been tested both in reptiles and mammals, while calcitonin of mammalian origin has been tested in reptiles, including species of turtles, lizards and snakes (Clark, 1972). Reptilian ultimobranchial extracts cause significant hypocalcaemic responses in rats similar to those of porcine or codfish calcitonin, but they are without effect upon serum Ca and phosphate levels in reptiles (Copp et al., 1972).

The effect of chronic administration of salmon calcitonin (4 μg kg^{-1} day^{-1})

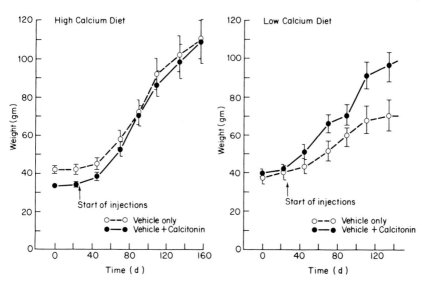

FIG. 48. Effect of chronic administration of salmon calcitonin (4 μg kg^{-1} d^{-1}) to hatchling turtles on a high Ca diet (left) and a low Ca regimen (right). (From Copp et al., 1972.)

to young turtles was investigated by Copp et al. (1972). While the hormone had no effect on growth in turtles fed high Ca diets, it significantly reversed the retardation of growth produced by a low Ca diet (see Fig. 48).

Histological examination of the tibial bones of the high Ca groups of reptiles revealed differences between control and calcitonin treated animals, the cortical bone appearing denser and the lacunae and canaliculi being reduced in size. These changes are consistent with an inhibitory effect of the hormone on osteolytic bone resorption (see also Bélanger et al., 1973).

Vitamin D_3

Reptiles apparently have the ability to synthesize the active vitamin D_3 metabolite 1,25-dihydroxycholecalciferol. This pathway has been demonstrated in lizards and turtles although not positively in the Garter snake, *Thamnophis elegans vagrans*, (Henry and Norman, 1975). Presumably the metabolite plays a role in the Ca metabolism of reptiles, but there appear to be no studies concerning the physiological effects of either the native vitamin or its metabolites in this class. It is to be hoped that this deficit in the knowledge of reptilian Ca metabolism will be remedied soon.

Other Hormones

There are no studies of hormones other than those mentioned above and those concerning gonadal hormones, in the reproducing reptile.

Reptilian Reproduction and Calcium Regulation

Calcium metabolism in reptilian reproduction has been meagrely studied compared with that in birds. There are a number of similarities between these classes with regard to production of a megalecithal egg, which, in reptiles, can have a more or less calcified eggshell. There are also some important points of contrast between these classes and Simkiss (1967) has reviewed this subject in detail. Since then, few papers on this topic have been published and only the main points of interests are summarized below.

Hypercalcaemia

Reproductively active female reptiles exhibit the hypercalcaemia typically associated with vertebrate vitellogenesis, these changes being particularly marked in snakes. Plasma Ca levels in the ribbon snake, for example, show a sixteenfold increase at the time of ovulation (see Table XXIX). Plasma

TABLE XXIX

Stage of reproductive cycle in the ribbon snake (*Thamnophis sauritus*)

	Inactive stage	Hydration stage	Deuto plasmic stage	Near ovulation	Early gravid
Ca (mmol litre^{-1})	2·74	3·35	7·32	43·50	3·61
Mg (mmol litre^{-1})	1·96	1·54	4·06	8·98	2·23
Pi (mmol litre^{-1})	1·64	1·50	2·41	7·00	2·31
Total protein g 100 ml^{-1}	4·06	4·22	4·95	6·67	3·91

(Modified from Dessauer and Fox, 1959.)

inorganic phosphate and magnesium also increase markedly at this time, reflecting the greater quantities of yolk protein precursors circulating in the blood.

The ribbon snake is, perhaps, not a typical reptile since it produces a large number of eggs with yolks very rich in Ca. Simkiss (1967) has estimated the Ca metabolism of the female to be 100 μmol kg^{-1} body wt h^{-1} during the reproductive season, a rate similar to that of a champion cow.

FIG. 49. The endolymphatic sacs of the female Philippine house lizard, *C. platurus* (right), are enormously enlarged structures filled with calcareous material, lying in the neck region. An X-radiograph of a male house lizard is shown for comparison (left). (From Simkiss, 1967.)

CALCIUM RESERVOIRS

Reptiles do not generally have such obvious Ca reservoirs (other than cortical bone) to compare with the massively developed endolymphatic $CaCO_3$ deposits of amphibia or the medullary bone of birds. In a few species of lizard, notably geckos and a few iguanids, extra-cranial endolymphatic deposits are relatively well developed: in some species, for instance the lizard *Cosmybotus platurus*, the sacs of females become engorged with $CaCO_3$ deposits at the time of reproduction (see Fig. 49). Shortly before oviposition, at a time when the eggs are fully formed with a calcified eggshell, the endolymphatic sacs are depleted of $CaCO_3$ (Ruth, 1918). Nothing is known of the mechanisms by which Ca is deposited or withdrawn from the sacs although, presumably, these processes are under the control of sex hormones. Whether or not parathyroid hormone is as important here as it appears to be in controlling amphibian endolymphatic Ca deposits, remains to be seen. Vitamin D_3 and its metabolites, as well as calcitonin, would also be worth studying in this respect.

Other reptiles, notably the Chelonia, exhibit changes in bone density associated with female reproductive activity. At the time of eggshell calcification, bone density in these animals falls sharply, while during the same period bone density in males remains constant (Edgren, 1960). Here again, there is considerable opportunity for further study of the hormonal influences affecting Chelonian bone metabolism in the reproducing female.

13. Calcium Regulation in Birds

Birds are the most widely studied of sub-mammalian vertebrates with respect to their calcium (Ca) metabolism. This partly reflects the economic importance of eggshell calcification in chickens which are bred for a high rate of egg production. Intensive egg production is often complicated by an inability of the hen's Ca metabolism to keep up, resulting in poor quality eggshells, a problem which can mean an economic loss in Britain alone of the order of several million pounds annually.

Birds, particularly chickens, are also widely used for laboratory work since their domestication ensures a cheap, plentiful and homogeneous supply. Owing to the importance and hence the richness of literature on Ca metabolism in the egg-laying hen, a large part of this chapter is devoted to that topic. First, however, Ca regulation in the non-reproducing or normal bird is reviewed.

Calcium Regulation in Normal Birds

Plasma calcium levels in birds other than egg-laying hens, are much the same as those in mammals (Simkiss, 1967). In chickens, for instance, they are typically about 2·05 mmol litre^{-1}. Birds have much the same complement of Ca regulating hormones and target organs as found in mammals; these include parathyroid hormone, calcitonin and vitamin D_3 metabolites as the major hormones, with bone, gut and kidney as the major target organs. Furthermore, since birds, like mammals, are homeotherms, we might expect the Ca metabolism of these two classes to be similar; in fact, there are some important differences.

PARATHYROID HORMONE

Anatomy

The parathyroids in birds are discrete, paired glands, situated in a position slightly caudal to the thyroids (see Fig. 14). There are usually four glands in the embryo which fuse in pairs on each side to give one glandular mass. Each parathyroid consists of cords of parenchyma cells as in mammals, although no oxyphil cells are present. The parenchyma (chief) cells have well-developed Golgi apparatus and large numbers of small secretory granules but few large storage granules (Taylor, 1971).

Parathyroid Extirpation

As in other classes of sub-mammalian tetrapods, the effect of parathyroidectomy in birds can be rather variable. Some birds exhibit only a slight hypocalcaemia which is rapidly corrected, while others show more severe symptoms including tetany and even death (Polin and Sturkie, 1958). It seems reasonable to say that these results reflect the presence of variable amounts of accessory parathyroid tissue, diffusely located, for instance, in association with ultimobranchial tissue (Dudley, 1942).

Parathyroid Hormone Injection

Birds react to parathyroid hormone injections by a transient rise in the plasma Ca level. The hypercalcaemic response to 100 U.S.P. U of bovine parathyroid hormone in chickens reaches a peak about three hours after subcutaneous injection, this effect being much larger in laying hens than in cockerels (Polin et al., 1957). More recently, attention has focused on the very short-term hypercalcaemic response to the hormone in birds. Candlish and Taylor (1970) reported that a low dose (20 U.S.P. U) of parathyroid hormone into hens gave rise to a transient hypercalcaemia within 10 min of i.v. injection. Immature birds also show a rapid and sensitive reaction to this hormone; week-old chickens or three-week-old Japanese quail respond to 2 U.S.P. U of i.v. parathyroid hormone within 15 to 60 min of injection (Dacke and Kenny, 1973; Kenny and Dacke, 1974). This response is widely used as the basis for a parathyroid hormone bioassay method (see Chapter 5).

Kenny and Dacke (1974) also investigated the physiological basis of the rapid parathyroid hormone response in immature Japanese quail; these studies were carried out mainly with the synthetic active fragment (1–34 peptide) of the bovine hormone. Intravenous injections of this material were given at the same time as an acute ^{45}Ca label (incorporated into the injection) and changes in total plasma Ca, as well as plasma ^{45}Ca, were monitored over periods of one or two hours. Data from this type of experiment for Japanese quail are shown in Fig. 50; chickens show similar responses. In both species, a marked hypercalcaemic response could be seen within 15 min of parathyroid hormone injection, ^{45}Ca levels being similarly elevated at this time. By 60 mins, however, radiocalcium levels had declined towards those of controls, although total plasma Ca levels were still significantly elevated. These responses were essentially similar whether the birds were Ca loaded at the time of hormone injection (as in the parathyroid hormone bioassay method of Dacke and Kenny, 1973), or normal (see Fig. 50). The difference between ^{45}Ca and total

158 CALCIUM REGULATION IN SUB-MAMMALIAN VERTEBRATES

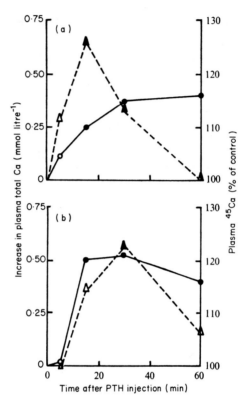

FIG. 50. Plasma total Ca and ^{45}Ca responses in 3-week-old Japanese quail after an i.v. injection of bovine (1-34) parathyroid hormone (PTH) containing ^{45}Ca such that each bird received approx. 2 μCi ^{45}Ca simultaneously with the PTH. (a) Non-Ca loaded quail receiving 10U PTH per bird. (b) Ca-loaded quail receiving 20U PTH per bird and 15 μmol CaCl$_2$ per bird simultaneously. Plasma total Ca results are expressed in terms of the increase of the mean PTH-treated levels over the control levels at each time interval. Plasma ^{45}Ca results are presented as the mean of the PTH-treated group expressed as a percentage of the control group for each interval. Those points indicated by (●) or (▲) were significant responses; ○——●, plasma total Ca; △——▲, plasma ^{45}Ca. A significant rise in plasma ^{45}Ca levels was seen at 15 and 30 min in (a) and at 30 min in (b). (From Kenny and Dacke, 1974.)

plasma Ca responses indicate at least two underlying mechanisms. The initial radiocalcium response suggests inhibition of Ca exit from the extracellular fluid; this effect could be mediated by kidney, gut or bone. The kidney seems an unlikely target organ since Buchanan (1961) and Candlish (1970) have both demonstrated hypercalciuric responses to parathyroid hormone in chickens, an opposite effect from that necessary to explain the ^{45}Ca response in quail and chicks. Similarly, the gut does not appeal as a target organ for the rapid parathyroid hormone response. There is little evidence in the literature to suggest a rapid, direct action of this hormone on gut, and the vitamin D_3 mediated effect of parathyroid hormone, which requires several hours to develop, is unlikely to account for the rapid response. Kenny and Dacke (1974) have suggested an inhibition of bone accretion as a feasible explanation. Such a response has also been implicated as the mechanism of the primary hypercalcaemic response to parathyroid hormone in rats, which occurs at about 90 min (Milhaud et al., 1971). The secondary response in avian total plasma Ca levels, occurring in this class at about 60 min and which is accompanied by a fall in plasma radiocalcium, can be accounted for in terms of bone resorption, i.e. the classical effect of parathyroid hormone.

Most of the work concerning the effect of parathyroid hormone on avian bone histology has been carried out with respect to the egg-laying bird and is considered in the section below which deals with that topic.

The main effect of parathyroid hormone on renal function in birds is to increase phosphate excretion (Levinsky and Davidson, 1975), a response which is mediated by a decrease in tubular reabsorption of this ion (Martindale, 1969). It also causes an instantaneous diuresis and an increase in Ca hydroxyproline and uronic acid excretion within 12 min of injection (Candlish, 1970). The doses of hormone used in these experiments were rather large (up to 125 U.S.P. U kg^{-1}) and lower doses (5 U.S.P. U kg^{-1}) gave only the phosphaturic response (Buchanan, 1961). The effects on phosphate, Ca and water excretion probably result from a direct action of the hormone on kidney, while the increased excretion of hydroxyproline and uronic acid reflect its action in mobilizing bone.

The adenyl cyclase system in crude plasma membranes of chick kidney seems to be ten to twenty times more sensitive to mammalian parathyroid hormone than a similar preparation from rat kidney cortex, perhaps partially accounting for sensitivity of avian species compared with mammals, to this hormone (Martin et al., 1974).

CALCITONIN AND THE ULTIMOBRANCHIALS

Anatomy

The C cells responsible for secretion of calcitonin in birds arise from the endoderm of the sixth branchial pouch in the embryo. These are located in the neck region, mainly as discrete, paired organs, the ultimobranchial glands, lying just caudal to the parathyroids and thyroids (see Fig. 14). There are also variable amounts of calcitonin secreting C cells located within other organs such as the thyroid or thymus (Simkiss and Dacke, 1971); however, as the authors pointed out, the ultimobranchial body in birds is a plentiful source of the hormone. In the fowl, the glands are richly vascularized structures about 2–3 mm in diameter. Hodges (1970) has reviewed in detail the histological structure of avian ultimobranchial glands.

Ultimobranchial Extirpation

Acute responses to ultimobranchialectomy in young turkeys were studied by Copp et al. (1970) during simultaneous infusion with Ca as $CaCl_2$ (10 mg Ca kg^{-1} h^{-1}). Whereas control birds were able to correct for the Ca load by an increase in calcitonin secretion, ultimobranchialectomized birds could not, but instead, showed a marked increase in the plasma Ca level following removal of the glands (see Fig. 51). Brown et al. (1970), studied the chronic response to ultimobranchialectomy in newly hatched chicks; five weeks after surgery, the workers were unable to detect any significant differences from control values of plasma Ca, alkaline phosphatase, or in bone composition. These chicks were unable to control a hypercalcaemic challenge induced by parathyroid hormone as efficiently as sham-operated controls. In further studies, the ultimobranchialectomized chicks were raised to maturity and laid eggs for several months, although these eggs tended to be smaller and thinner shelled than normal (Speers et al., 1970).

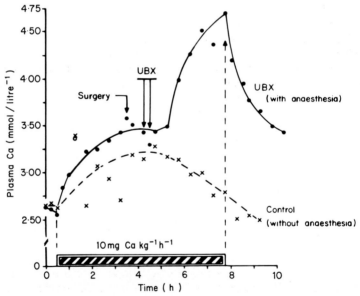

FIG. 51. Effect of bilateral ultimobranchialectomy performed during the infusion of a young turkey with 10 mg Ca (as $CaCl_2$) kg^{-1} h^{-1}. (From Copp et al., 1970.)

Calcitonin Injection

Many workers have attempted to demonstrate acute hypocalcaemic responses to injected calcitonin in birds, but without success (Dacke, 1972; Urist, 1967;

CALCIUM REGULATION IN BIRDS 161

Candlish and Taylor, 1969). Other workers (Kraintz and Intscher, 1969; Calamy and Barlet, 1970), claim to have produced positive responses in cockerels and chicks respectively, although their data are lacking in statistical quantitation and have yet to be repeated under properly controlled conditions.

Chronic responses to prolonged calcitonin treatment in both immature and mature male and female chickens have been investigated by Bélanger and Copp (1972). In this study, salmon calcitonin was administered daily as a subcutaneous dose of 2 MRC U, under a variety of experimental conditions. In young (one-month-old) birds, the hormone caused a decrease in osteocytic and osteoclastic resorption after four months of treatment, but was without any such effect after only one month of treatment in older (two or three month) male or female birds. The latter groups of birds did increase their weight, however, compared with controls, by 20% in the case of the males and 11% in females. The bones of the treated birds appeared denser and more mature than controls, probably as a result of decreased bone resorption, these responses being consistent with those reported in turtles by Copp et al. (1972). Similar increases in bone weight (25%) and alkaline phosphatase activity (70%) were found in chick embryos following chronic treatment with salmon calciton (500 MRC mU d^{-1} embryo^{-1}), but a slightly higher dose (750 MRC mU d^{-1}) had small inhibitory effects on these parameters (McWhinnie, 1975) which were suggested by the author to be a result of a secondary hyperparathyroidism. Since an inhibition of bone resorption is considered to be the principal mechanism of the hypocalcaemic response to calcitonin in mammals, it is surprising that the analogous bone response in birds is not accompanied by hypocalcaemia. One possible explanation of this is that any slight hypocalcaemia is immediately corrected by a rapidly acting endogenous parathyroid hormone. Chronic responses of bone to calcitonin in egg-laying birds are discussed in a later section of this chapter.

Calcitonin Secretion

Much of the information concerning the mechanism of calcitonin secretion in birds and mammals has been summarized by Care and Bates (1972). While it is clear that secretion of this hormone from the avian ultimobranchial is stimulated by hypercalcaemia, it is also apparent that the gland, at least in geese, is less responsive to a given increase in plasma Ca concentration than is the mammalian thyroid (see Fig. 52). However, the resting level for calcitonin secretion in the goose is much higher than that reported for the pig and sheep (Bates et al., 1969); thus the rate for the goose at around 2 MRC mU (min^{-1} kg^{-1} body wt), is twenty-five times that found in pigs and sheep. This finding is also supported by the evidence from Boelkins and Kenny (1973), who reported very high plasma calcitonin levels in birds, particularly in Japanese quail (see Chapter 5).

In 1970, Ziegler et al. demonstrated that cyclic-AMP can stimulate calcitonin release in birds. Only β-adrenergic compounds such as isoprotorenol

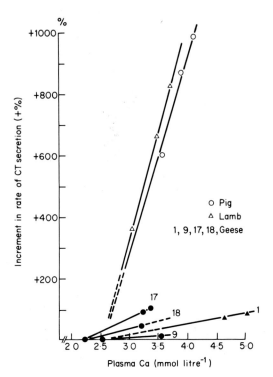

FIG. 52. Comparison between the percentage increase in calciton in secretion rate vs plasma Ca concentration in the blood used to perfuse the ultimobranchial gland of the goose or thyroid gland of the pig and lamb. Doubling the plasma Ca concentration caused a much greater increase in calcitonin secretion from porcine and ovine thyroid than from anserine ultimobranchial gland. (From Care and Bates, 1972.)

have so far been shown to activate an avian C cell adenyl cyclase: this may be significant when we consider the innervation of the ultimobranchial glands which is of a more complex nature than in the equivalent mammalian tissue (Hodges, 1970). Vagal stimulation apparently results in a two- or threefold increase in the rate of calcitonin secretion in birds (cited in Care and Bates, 1972) leading to the suggestion by these authors that the control of calcitonin secretion in this class may be largely neural rather than humoural.

Calcitonin secretion has been studied in vitro in cultures of chicken ultimobranchial glands by Feinblatt and Raisz (1972) and more recently by Nieto et al. (1975), the tissue being particularly suitable for such studies due to the high degree of cell homogeneity. Calcitonin secretion by these cultures is stimulated by increasing the Ca concentration of the medium and also by application of dibutyryl cyclic-AMP at a concentration of 2×10^{-3} mol litre^{-1}.

Apart from the influences of Ca ions on calcitonin secretion, there is evidence that intestinal factors play a role in the release of this hormone; these have been discussed in Chapter 5.

Vitamin D_3 Metabolites

Chemistry

Avian species, particularly chickens, have made a major contribution towards our general knowledge of the metabolism and physiological function of vitamin D_3. Birds possess enzyme systems necessary for the hydroxylation of cholecalciferol to 25-hydroxycholecalciferol in the liver (Omdahl and De Luca, 1973) and subsequent metabolism of this substance to 1,25-dihydroxycholecalciferol in the kidney (Henry and Norman, 1975), who also showed the kidney 25-hydroxycholecalciferol-1-hydroxylase to be almost four times more active in rachitic chicks (520 pmol min^{-1} g^{-1} kidney) than in those given diets supplemented with vitamin D_3 (140 pmol min^{-1} g^{-1} kidney).

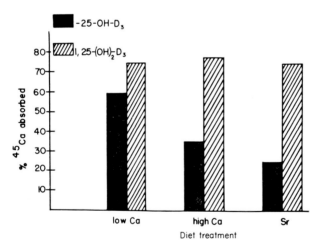

Fig. 53. Influence of 25-hydroxycholecalciferol (25-OHD$_3$) and 1,25-dihydroxycholecalciferol (1,25-(OH)$_2$D$_3$) on intestinal Ca absorption in chicks fed a low (0·08%) or high (2%) Ca or strontium (2·35%) diet. Chicks were fed a vitamin D deficient diet until they were 2 weeks old; they were then divided into 3 groups and fed indicated diets. At the same time half of each group received 25-OHD$_3$ (130 pmol) and the other half 1,25-(OH)$_2$D$_3$ (130 pmol), given daily as an oral dose. Calcium absorption was measured at 4 days (strontium group) or 11 days (low and high calcium groups) after initiation of dietary and vitamin treatment. Each value is mean of 5 observations. (From Omdahl and De Luca, 1973.)

Effects

The active vitamin D_3 metabolites have similar physiological effects in birds to those in mammals. Thus 1,25-dihydroxycholecalciferol affects gene transcription in the avian intestine which, in turn, initiates the formation of calcium binding protein (Wasserman and Corradino, 1971; Omdahl and De Luca, 1973). These reviews also give details of the effect of the metabolite on avian bone, which is greatly to increase mobilization of the mineral reserve. The differential effects of 25-hydroxycholecalciferol and 1,25-dihydroxycholecalciferol on intestinal Ca absorption in chicks fed high or low Ca or strontium diets are shown in Fig. 53. Chicks fed low Ca diets showed an increased effect of 25-hydroxycholecalciferol on Ca uptake, reflecting the increased production of 1,25-dihydroxycholecalciferol in the low Ca situation.

OTHER HORMONES

Apart from the three major systems discussed above, the most influential of the hormones affecting Ca metabolism in birds are probably the gonadal steroids; these will be discussed below in relation to avian reproduction. Target organs in avian Ca metabolism appear to be essentially similar to those in mammals, with the exception of medullary bone formation in hens during the reproductive season. Thus in the non-reproducing bird, Ca regulation has some similarities to the reptilian system and some to the mammalian system. It would be imprudent, however, to take this as evidence that birds merely represent a transitional stage between the other two classes in respect of their Ca regulation. Apart from interesting adaptations associated with eggshell formation, their Ca metabolism may have become adapted in relation to the modification of bone structure for the purpose of flight. Another neglected area concerns the divalent ion metabolism in sea birds.

Calcium Metabolism in Egg-laying Birds

Calcium regulation in egg-laying birds has attracted a great deal of attention for more than 40 years, largely because of the economic problem associated with thin eggshells. Eggshell quality has decreased over the past few years: in the U.K. for instance, in 1960 about 5% of eggs were downgraded due to shell damage whereas by 1970 the figure had reached more than 6%. This may be partly a result of modern cage design and husbandry (Carter, 1971).

Several reviews have been published over the last few years dealing with various aspects of Ca metabolism in egg-laying birds, reflecting the rapidity with which our knowledge of this subject is being augmented. One such

review by Simkiss (1975) covers this field and in it, he pays particular attention to the dynamic and regulatory aspects of Ca metabolism. He also summarizes knowledge of Ca metabolism in reproducing birds as it was about twelve years ago into six major points. These are:
(1) That the egg-laying bird can only absorb a limited supply of Ca across the intestine and that this necessitates some form of reservoir for use during periods of Ca stress.
(2) That 98% of Ca in birds is in the bone and that this store may be laid down or repleted in association with different periods of eggshell formation.
(3) That the hen's skeleton exists in two forms, i.e. cortical and medullary bone, and that it is the highly labile medullary bone which acts as a Ca reservoir for use by the egg-laying birds. Medullary bone requires the presence of oestrogens and androgens for its formation.
(4) Calcium metabolism in egg-laying birds is greatly affected by the vitamin D_3 status, but parathyroid hormone appears to be even more important in controlling plasma Ca levels by virtue of its effects on bone, gut and kidney.
(5) The "shell gland" in the oviduct of the egg-laying chicken is capable of secreting about 5 g of $CaCO_3$ in 20 h. The ions are removed from the blood during this process and carbonic anhydrase appears to be important in the provision of carbonate ions.

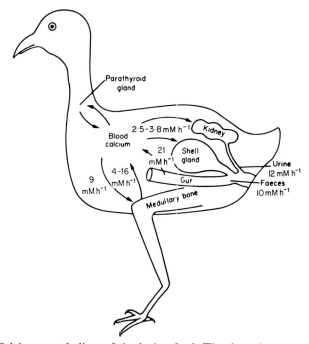

FIG. 54. Calcium metabolism of the laying fowl. The data give rates for most of the processes involved (From Simkiss, 1975.)

166   CALCIUM REGULATION IN SUB-MAMMALIAN VERTEBRATES

(6) The avian embryo is capable of resorbing about 5% of the eggshell during incubation and this provides approx. 80% of the Ca requirements of the embryo. Some of these ideas are summarized in Fig. 54.

Most of the concepts summarized above are founded on balance studies which Simkiss (1975) considers give a rather static picture of the Ca metabolism in the egg-laying bird. He suggests that recent studies have dealt more with dynamic concepts of Ca metabolism, particularly in terms of the cellular and membrane role in ion transfer, and also with the differential response times of the various regulatory systems.

In the discussion below, an attempt is made to summarize our knowledge of regulatory systems involved in the hen's Ca metabolism. A major question still facing us concerns mechanisms by which these systems act in concert to ensure the Ca balance of the egg-laying hen.

THE PARATHYROIDS IN EGGLAY

The role of the parathyroids in egg-laying hens was reviewed by Taylor (1971) and more recently by Kenny and Dacke (1975).

Changes in Size

In hens, parathyroid glands undergo enlargement with the onset of reproductive activity. Similar hypertrophic and hyperplastic responses are observed in these glands if birds are fed diets deficient in Ca or vitamin D_3, while diets relatively deficient in phosphorous (0·4%) and containing excess Ca (3·0%) cause atrophy of the glands in growing pullets (Taylor, 1971).

Parathyroid Hormone Injection

It is well known that egg-laying hens show more marked hypercalcaemic responses to parathyroid hormone than do cockerels (Polin et al., 1957; Urist et al., 1960) (see Fig. 17). Most of this enhanced response can be accounted for by a greater response in the protein bound fraction of blood Ca in the hen; the rise in diffusible Ca is similar for both sexes. It has been suggested that the greater sensitivity of hens to the hormone is due to increased binding of Ca to yolk proteins as the ionized Ca fraction rises (Simkiss, 1967; Taylor, 1971). An alternative argument is that the response in hens reflects the presence of additional target organ receptors such as medullary bone (Dacke, 1970; Kenny and Dacke, 1975). Another target organ which could account for the increased sensitivity of egg-laying hens to parathyroid hormone is the avian oviduct. Dacke (1976) produced indirect evidence to suggest that the hormone acts upon the oviduct to limit the secretion of Ca into the calcifying eggshell. In these experiments the effect of a single large injection of parathyroid hormone

(50 U.S.P. U per bird) into Japanese quail hens, 4 h after ovulation, resulted in a decrease in eggshell thickness expressed as shell wt per U area. In the same experiment the specific activity of a chronic (2 week) ^{45}Ca label was increased in the calcifying eggshell, indicating increased resorption of bone Ca which contributed a larger fraction to the shell Ca compared with that from dietary calcium (see Fig. 55). When hormone injection was delayed until

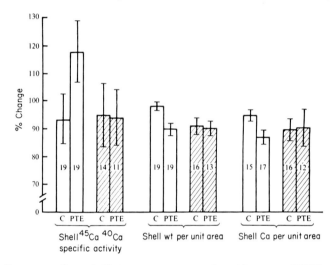

FIG. 55. Response of eggshell parameters to parathyroid extract (PTE) or control vehicle (C) injections; columns represent mean percent change between pre- and post-injection eggs ± s.e. Experiment 1, birds injected within 2–6 h of oviposition (unhatched columns). Experiment 2, birds injected within 12–16 h of oviposition (hatched columns). Numbers in columns indicate birds per group. (From Dacke, 1976.)

approx. 14 h after ovulation, it had none of the above effects. It was proposed that the disappearance of response in this latter experiment might simply have reflected high endogenous parathyroid hormone levels in the later part of the ovulation cycle, with subsequent saturation of target organ receptors by this hormone.

Evidence for Variations in Endogenous Parathyroid Hormone Activity

As yet there are no reports concerning the assay of circulating parathyroid hormone during the hen's ovulatory cycle. There is considerable circumstantial evidence, however, for a cyclical fluctuation in the hormone activity during different phases of this cycle. Thus periods of medullary bone resorption occur at times when no dietary Ca is likely to be available, but eggshell formation is active. Osteoclastic resorption of medullary bone occurs in pigeons during eggshell calcification, while at times when no eggshell is calcifying, osteoblastic activity is more pronounced. These changes are paralleled by gross fluctuations in the quantity of medullary bone (see Fig. 11). Other physiological changes consistent with bone resorption during eggshell formation include increases in urinary phosphate and hydroxyproline, while circulating acid and alkaline phosphatases also show variations consistent with different phases of bone turnover (Kenny and Dacke, 1975). Surprisingly, medullary bone seems to be more resistant to systemic Ca depletion than does cortical bone, so that in egg-laying hens maintained on a low Ca diet, the quantity of cortical bone is reduced by about 40% during six successive ovulation cycles, while medullary bone remains more or less constant (Taylor and Moore, 1954). It is fairly clear that primary formation of medullary bone is induced by the synergistic actions of oestrogens and androgens (while cortical bone is not) but it is unclear what factor or factors control its cyclical periods of resorption and replacement in the egg-laying hen. Taylor (1970) suggests that under normal circumstances medullary bone, by virtue of its high vascularity, large surface area and diffuse nature of its mineral phase, may be more sensitive to small changes in circulating parathyroid hormone than is cortical bone, but that under periods of dietary Ca deficiency, the levels of endogenous hormone are greatly elevated.

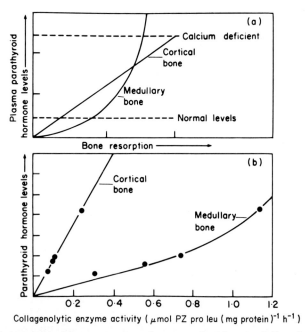

FIG. 56. (a) A graphical interpretation of a suggestion that medullary bone is more responsive than cortical bone to low levels of parathyroid hormone but less responsive at high levels. (b) Collagenolytic enzyme activities of cortical and medullary bone under various conditions. (From Simkiss, 1975.)

These would tend to have more profound effects on cortical rather than medullary bone which quickly achieves its maximal response to the hormone. Simkiss (1975) has summarized these ideas into a series of graphs and they are further substantiated by the results of Bannister and Candlish (1973) who measured collagenolytic activity in avian bone under the influence of parathyroid hormone (see Fig. 56). With low levels of the hormone, cortical bone appeared to be less sensitive than medullary bone, but there was some indication that as the hormone levels rose, medullary bone showed a plateau in its response, while cortical bone did not.

It seems likely that parathyroid hormone plays a differential role during the hen's ovulatory cycle by causing resorption of both cortical and medullary bone, although the medullary bone response may be dominant under normal conditions. The hormone presumably also induces gut Ca absorption indirectly by virtue of its effect on vitamin D_3 metabolism and may influence the kidney, mainly in terms of increased phosphate excretion. There is also evidence for parathyroid hormone having an effect, either direct or indirect, on the avian oviduct to limit $CaCO_3$ secretion into eggshell. The overall combined evidence suggests that the primary role of the hormone during eggshell formation is to protect the plasma Ca level, if necessary at the expense of bone and calcifying eggshell.

During the chicken's ovulatory cycle there are fairly marked fluctuations in plasma Ca levels: as eggshell formation proceeds there is a fall in total plasma Ca which is reflected in the diffusible or ionized Ca level (Taylor and Hertelendy, 1961; Hertelendy and Taylor, 1961). Similar changes occur in the total and ionized Ca levels of egg-laying Japanese quail (Dacke *et al.*, 1973). Presumably the progressive hypocalcaemia associated with eggshell formation acts as a stimulus for parathyroid hormone secretion which, in turn, affects organs such as bone and possibly oviduct to limit the fall in plasma Ca.

There is also preliminary evidence for a parathyroid influence on the Ca appetite in birds which is discussed below.

CALCITONIN IN EGGLAY

The role of calcitonin in egg-laying birds is less well understood than that of parathyroid hormone.

Glandular and Plasma Calcitonin Levels

The ultimobranchials, like the parathyroids, hypertrophy in egg-laying hens (Urist, 1976). Levels of glandular calcitonin in hens, however, are not higher than in males and furthermore, they exhibit a considerable decline when compared with levels in young birds (see Table XXX). Somewhat surprisingly, plasma calcitonin levels (at least in Japanese quail, a species in which they have been extensively studied by Kenny and his coworkers), do not show the same

TABLE XXX
Levels of plasma Ca, weight of ultimobranchial glands and calcitonin content of domestic fowl of different ages

Age (day)		Plasma (mmol litre^{-1})	Wet weight ultimobranchial glands (mg)	Calcitonin content per bird (MRC mU)	Calcitonin kg^{-1} body wt (MRC mU kg^{-1})
Embryo	18	—	0·5	42	2500
Hatched	3	2·25	0·4	163	5200
	35	2·48	3·4	595	2000
	77	2·50	5·0	1085	1400
	105	2·48	6·4	1351	1000
	490 ♂	2·70	5·6	1495	600
	490 ♀	5·25	5·3	1060	600

(From Dent et al., 1969.)

patterns as the glandular levels of chickens. Thus while these levels are similar in immature quail and egg-laying females, in adult males they are about three times higher than in females (Boelkins and Kenny, 1973). During the ovulation cycle there are modest fluctuations in the plasma calcitonin levels, these being highest about 7 h after ovulation and falling by about 37% during eggshell calcification (Dacke et al., 1972). As might be expected, the hormone levels are inversely correlated with total plasma Ca levels in egg-laying quail. Another interesting feature of plasma calcitonin levels in female quail is that they show a surge followed by a fall, about one or two weeks before commencing ovulation (see Fig. 57). This surge occurs shortly before the onset of hypercalcaemia associated with egglay, suggesting that the change in plasma Ca level is not the primary stimulus for the hormone surge (Dacke et al., 1976). Similar surges of plasma calcitonin are found in pre-spawning salmon and rutting stags (see Chapters 5 and 10). Dacke et al. (1976) proposed that gonadal hormones act in some way to modulate plasma calcitonin levels since these are lowered in males following castration. Shortly before the commence of egglay, at least in chickens, a surge in plasma levels of gonadal steroids (both oestrogens and androgens) occurs (Senior, 1973), and Dacke et al. (1976) suggested that these act as a stimulus for the pre-laying surge in calcitonin secretion.

Calcitonin Injection

In addition to their studies of the effects of chronic calcitonin injection on bone metabolism in non-laying birds, Bélanger and Copp (1972) studied the effects of this hormone in egg-laying hens on high Ca diets. These birds received an intramuscular dosage (2 MRC U per day) of the hormone for four months. The total amount of medullary bone was decreased by this treatment,

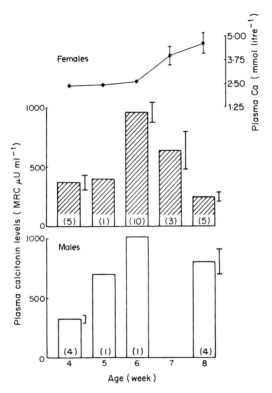

FIG. 57. Changes in plasma calcitonin (CT) levels in female quail (hatched columns) and male quail (unhatched columns) shortly before maturity ($\pm$ s.e.). Numbers in columns represent the number of pooled samples assayed for CT activity. The plasma Ca levels of the female quail are plotted at the top of the figure ($\pm$ s.e.). The plasma Ca of females did not rise significantly until 7-weeks-old, after the surge in plasma CT levels. By 8 weeks the plasma Ca levels were continuing to rise as the birds commenced egg-laying whereas the plasma CT levels had fallen back to pre-surge levels. The rise in plasma CT of females is significant ($p < 0.01$) at 6 weeks compared with 4-week levels. By 8 weeks the levels are significantly ($p < 0.01$) decreased compared with the 6-week levels. (From Dacke et al., 1976.)

but the cortical bone became denser, although it also contained several large resorption cavities, indicating increased remodelling. The authors concluded that calcitonin does nothing to protect the egg-laying birds against Ca stress and indeed may contribute towards it by lowering plasma Ca levels. One possibility is that the bones were exhibiting "escape" from this hormone (see Chapter 6) and then responding to endogenous parathyroid hormone, the levels of which probably fluctuate during the hen's ovulatory cycle (see above).

The only published study of the influence of calcitonin on eggshell formation appears to be that of Joshua (cited by Mueller and Leach, 1974). He found that intramuscular injection of porcine calcitonin (10 MRC U kg^{-1}) at the onset of eggshell calcification had no effect on eggshell thickness. Similarly, I failed to find any effect of salmon calcitonin in Japanese quail hens (0·3 MRC U 100 g^{-1} i.p.) on eggshell thickness, whether injected within about 6 or 14 h after ovulation (Dacke, unpublished observations). There does not seem to be any evidence at present to indicate an effect of the hormone, whether inhibitory or stimulatory, on secretion of Ca by the avian oviduct.

The physiological role of calcitonin in egg-laying birds then, remains obscure. Perhaps its function is to prevent hypercalcaemic overshoot by the fast acting parathyroid system at the end of eggshell calcification. An alternative role, which may be to protect bone against excessive parathyroid hormone induced resorption, (see Chapter 6), is not supported by the evidence presently available.

Vitamin D_3 in Egglay

Deficiency

A dietary deficiency of vitamin D_3 in egg-laying hens results in oesteomalacia, dwindling egg production and a deterioration in eggshell quality. These effects are attributed to poor absorption of Ca from the gut and disturbances in bone turnover (Coates, 1971). Undoubtedly, therefore, the vitamin plays an important role in maintaining the Ca balance of egg-laying birds.

Metabolism

Recently, in an elegant series of studies, Kenny and his associates have demonstrated that vitamin D_3 metabolism in the hen is considerably influenced by the ovulation cycle, the principal metabolic end product appearing to depend on whether or not the hen had ovulated. In hens which had ovulated 1–6 h previously, 69% of the metabolic end product was represented by 1,25-dihydroxycholecalciferol, with the rest as 24,25-dihydroxycholecalciferol. The quantity of 1,25-dihydroxycholecalciferol produced dropped slightly to 59% of the total by 18–24 h after ovulation. In hens which followed an oviposition by a clutch pause, the quantity of 1,25-dihydroxycholecalciferol produced started to fall about 3 h after the oviposition, eventually reaching about 20% of the total (Kenny, 1974, 1976). These data are summarized in Fig. 58. They have been confirmed and extended by Sedrani and Taylor (1977) who measured 1,25-dihydroxycholecalciferol production in egg-laying quail which had either ovulated, or had not ovulated for one to three days, as well as in non-laying hens. Their results are summarized in Table XXXI. While Kenny (1974) considers that ovulation in some way provides a direct trigger for 1,25-dihydroxycholecalciferol synthesis, Sedrani and Taylor (1977) felt that para-

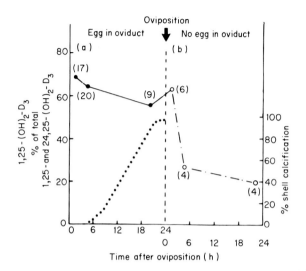

FIG. 58. *In vitro* renal production of 1,25-dihydroxycholecalciferol (1,25-$(OH)_2$-D_3) expressed as a percent of total 1,25- and 24,25-$(OH)_2$-D_3 biosynthesis, at different periods (1–3, 3–6 and 18–24 h) after oviposition in Japanese quail with (●——●) and without (○— - —○) an egg in oviduct. Ovulation, when it occurs, is within 30 min after oviposition. The production of 1,25-$(OH)_2$-D_3 is already enhanced (69% and 64%) at 1–3 and 3–6 h after oviposition/ovulation (a) at a time before calcification of eggshell has begun, and remains enhanced throughout 24 h ovulatory cycle. If oviposition is not followed immediately by ovulation (b) then 1,25-$(OH)_2$-D_3 production, although still enhanced 1–3 h after oviposition, falls rapidly and significantly ($p < 0.05$) to 27% and 20% at 3–6 and 18–24 h after oviposition, respectively. Data taken from Woodard and Mather (1964) indicating eggshell calcification rate (..........). Numbers in parentheses represent number of birds contributing data to mean value. (From Kenny, 1976.)

TABLE XXXI

In vitro production of 1,25-dihydroxycholecalciferol ($(OH)_2D_3$) as % total 1,25-plus 24,25-($(OH)_2D_3$) by kidney homogenates of female Japanese quail in a variety of physiological states (values are mean ± s.e.)

Physiological state	No of birds	Mean wt of ovary (g)	1,25 $(OH)_2D_3$ production % of total
No egg in oviduct			
Non-laying	7	0·76 ± 0·34	22 ± 5·8
1-day pause	6	6·45 ± 0·94	70 ± 6·8
3-day pause	4	6·08 ± 0·82	0
1 egg in oviduct	7	4·97 ± 0·56	85 ± 5·8

(From Sedrani and Taylor, 1977.)

thyroid hormone might provide the stimulus since levels of this hormone are probably higher on egg-laying days due to the decline in plasma Ca levels, than on non-laying days, and might, therefore, provide a trophic stimulus for 25-hydroxycholecalciferol-1-hydroxylase activity (see Chapter 5). Nevertheless, Baksi and Kenny (1976a) have demonstrated that ovariectomized quail hens lose the capacity to synthesize the 1,25-metabolite although this can be restored by providing a low Ca diet. Furthermore, these workers showed that chronic treatment of female quail with testosterone reduces the capacity of these birds to produce 1,25-dihydroxycholecalciferol even though plasma Ca levels are lowered, while oestradiol treatment of males not only causes hypercalcaemia but also increases their capacity to synthesize 1,25-dihydroxycholecalciferol (Baksi and Kenny, 1976b). Treatment of quail hens with an anti-oestrogen (ICI 46 474) at a dose of 30 mg kg^{-1} intramuscularly three times per week for four weeks, completely abolished their capacity to synthesize either the 1,25 or 24,25-metabolite (Baksi and Kenny, 1976c). It is clear then, that in addition to any trophic influence of parathyroid hormone or plasma Ca level, production of the active vitamin D_3 metabolite, in birds at least, is greatly influenced by sexual status. In this contexture oestrogens stimulate while androgens inhibit production of the metabolite. Similar data concerning a stimulatory action of oestrogen on 1,25-dihydroxycholecalciferol production in rats have been obtained by Baksi and Kenny (1977). It remains to be seen whether oestrogens and androgens act directly on the kidney 25-hydroxycholecalciferol-hydroxylases or whether their influences are mediated by differential secretion of prolactin which, in turn, modulates vitamin D_3 metabolism (see Chapter 5).

Obviously it is useful for an egg-laying bird to produce 1,25-dihydroxycholecalciferol on days when an egg is present in the oviduct in order to facilitate gut Ca absorption and presumably bone resorption. We have also seen that 24,25-dihydroxycholecalciferol is active, albeit at a much lower level than 1,25-dihydroxycholecalciferol, with respect to intestinal Ca absorption, but not bone resorption (see Chapter 6). This presumably enables egg-laying birds to at least partially replenish their bone Ca reservoirs during pauses between clutches of eggs. The question remains as to whether this replenishment could be accomplished more efficiently if the principal metabolite in the circulation on non-laying days were to be 1,25-dihydroxycholecalciferol rather than 24,25-dihydroxycholecalciferol. If it were possible in commercially important species such as chickens, to produce a more continuous supply of the 1,25-metabolite, the Ca reservoirs might be replenished more readily, leading indirectly to an improvement in eggshell quality.

Effect of Photoperiod

A possible approach to this problem might be to regulate the interval between successive ovulations in hens and cut down the number or duration of clutch pauses. Since ovulation in birds is regulated by photoperiod, this can be accomplished by increasing the interval between ovulation and oviposition through use of ahemeral light cycles (i.e. other than 24 h). This has been tried with chickens (Morris, 1973) and Japanese quail (Dacke, 1977). In both cases light cycles of around 28 h lead to increases in eggshell thickness of between 7 and 10% over experimental periods lasting several months. In quail, at least, the increase is not accomplished at the expense of the skeleton, indicating that a compensatory increase in dietary Ca uptake occurs (Dacke, unpublished observation).

Effect on Gut

The rate of Ca absorption across the gut is very high in egg-laying birds and can be measured by using an unabsorbed marker (^{91}Y) to trace (^{45}Ca) absorption along different parts of the gut. Hurwitz et al. (1973), using this technique, demonstrated a general increase in Ca absorption along the gut when eggshell formation is occurring (Table XXXII). Presumably these responses are related

TABLE XXXII
Percentage absorption of Ca in different intestinal segments of chickens in different physiological states

	Duodenum	Upper jejunum	Lower jejunum	Upper ileum	Lower ileum
Laying hens—no shell	18·6	17·7	2·3	0·8	−3·4
Laying hens—shell formation	25·1	45·7	6·0	6·9	−2·1
Laying hens—Ca depleted	53·5	12·5	4·8	1·6	1·0
Non-laying pullets	−11·6	40·1	4·1	1·8	−11·4

(After Hurwitz et al., 1973.)

to the high levels of 1,25-dihydroxycholecalciferol production in ovulating hens. When the hens are severely Ca depleted, a second type of adaptation occurs which appears to be restricted to the duodenum (see Table XXXII). This is reminiscent of the adaptation in frog gut following ultimobranchialectomy (see Chapter 11) and one might ask if calcitonin is also involved in the duodenal response of the calcium depleted bird, with the hormone normally having an inhibitory action on gut Ca absorption at this point.

Effect on Oviduct

There are some similarities in the mechanisms of Ca transport across the gut and oviduct in birds and it would, therefore, seem reasonable that the latter process might be influenced by vitamin D_3 metabolites. Much of the experimental work carried out with respect to Ca secretion by the avian oviduct has been done by the Schraers and their coworkers who have shown that the oviduct can transport Ca against a concentration gradient (Schraer and Schraer, 1965; Schraer *et al.*, 1965). Very little is known, however, concerning influences of Ca regulating hormones on this organ. The oviduct appears to contain a Ca binding protein similar to that found in gut, which may be sensitive to changes in vitamin D_3 status (Corradino *et al.*, 1968), although this is disputed since it does not appear to be increased during vitamin D_3 deprivation (Bar and Hurwitz, 1973). It is possible that the apparent inhibitory effect of parathyroid hormone on avian eggshell calcification (Dacke, 1976), discussed above, could be mediated indirectly by increased 1,25-dihydroxycholecalciferol synthesis which, in turn, inhibits Ca transport by this organ, though this would seem to be the opposite effect from what might be expected with the vitamin D_3 metabolite.

ACID-BASE BALANCE IN THE EGG-LAYING HEN

Eggshell Carbonate

The major anion to accompany Ca secretion into the eggshell is carbonate, which is derived from metabolic carbon dioxide produced by the shell gland rather than from plasma bicarbonate. Hydration of carbon dioxide to bicarbonate is probably catalysed by carbonic anhydrase, an enzyme which is present in the shell gland mucosa in significant quantities. Thus inhibitors of carbonic anhydrase, such as acetazolamide at a dose of 12 to 25 mg kg^{-1} body wt in hens, cause a reduction of eggshell weight of around 80% (Mueller and Leach, 1974). Since carbonic anhydrase appears to be involved in the action of parathyroid hormone in organs such as bone, kidney and amphibian endolymphatic sacs (see Chapters 6 and 11), it seems reasonable that the hormone should influence the enzyme in the avian oviduct. This influence is presumably an inhibitory one since the hormone also limits eggshell calcification. Such an inhibitory role would be consistent at least with the effect of parathyroid hormone on carbonic anhydrase in the mammalian nephron, which is also of an inhibitory nature (see Chapter 6).

It was suggested by Simkiss (1961) that the reaction involved in formation of carbonate ions in the avian oviduct could be represented by the equation:

$$H_2O + CO_2 \rightarrow CO_3^{2-} + 2H^- \qquad (1)$$

This hypothesis was supported by the experimental work of Lorcher and Hodges (1969) who injected $^{45}CaCl_2$ and $NaH^{14}CO_3$ simultaneously into the veins of laying and non-laying hens, blood being collected from the sciatic artery and also the vein draining the shell gland. Radioisotope analyses showed that while the ^{45}Ca was reduced by 57% during eggshell formation, ^{14}C levels in the blood stayed constant. No changes occurred in the isotope levels of the blood flowing through the shell gland in non-laying birds. These workers similarly demonstrated that during 2 h of shell formation 38–45% of the ^{45}Ca dose could be detected in the eggshell, while less than 3% of the ^{14}C dose was recovered. From these data it was concluded that the carbonate ion could not be derived directly from plasma bicarbonate according to the reaction:

$$2HCO_3^- \rightarrow H_2CO_3 + CO_3^{2-} \rightarrow \boxed{Eggshell}$$
$$\downarrow \qquad\qquad\qquad\qquad\qquad (2)$$
$$\boxed{\text{removed via carbonic anhydrase}}$$

but rather that the reaction suggested by Simkiss (1961) is correct.

Acidosis in Hens

It is assumed that in conjunction with secretion of carbonate ions into eggshell, protons are transferred into the blood. In a classic paper, Mongin and Lacassagne (1964) were able to demonstrate a cyclical metabolic acidosis in hens associated with formation of the eggshell; cockerels showed no such changes. Had the authors of this paper carried out control experiments with non-laying hens they might have given a different interpretation to their results: such controls were used by Dacke et al. (1973) in a similar series of experiments involving Japanese quail hens. While the egg-laying hens exhibited cyclical fluctuations in blood pH and PCO_2, so did the non-laying controls, but the males did not. Furthermore, the metabolic acidosis in the egg-laying group of quails reached a peak about half-way through eggshell calcification rather than towards the end as in chickens. Since chickens lay their eggs in the morning, while quails lay theirs in the afternoon, it was suggested by Dacke and his colleagues that the cyclical metabolic acidosis in hens has some circadian origin not directly associated with eggshell formation. To be fair to the proponents of the view that eggshell formation is accompanied by a secretion of protons into the bloodstream (and I am one of them), there is considerable experimental evidence to support such a theory (Simkiss, 1975).

While the equation suggested by Simkiss (1961) seems the most suitable one to explain the secretion of carbonate into an avian eggshell, it is not clear how carbonic anhydrase is involved in this system, although there is no doubt that it is. Perhaps one should rewrite Simkiss's equation as:

$$H_2O + CO_2 \xrightleftharpoons[\text{anhydrase}]{\text{carbonic}} H_2CO_3 \rightarrow \underset{\boxed{\text{Blood}}}{2H^+} + \underset{\boxed{\text{Eggshell}}}{CO_3^{2-}} \quad (3)$$

A further fact to be considered is the finding of Campbell and Speeg (1969) that deposition of $CaCO_3$ by the hen's oviduct and also in the land snail shell (a similar process) is associated with the formation of ammonia. They have suggested that this promotes the precipitation of $CaCO_3$ by titrating bicarbonate ions:

$$HCO_3^- + NH_3 \rightarrow CO_3^{2-} + NH_4^+ \quad (4)$$

Perhaps equation (3) should, therefore, be rewritten as:

$$H_2O + CO_2 \xrightleftharpoons[\text{anhydrase}]{\text{carbonic}} H_2CO_3 \rightleftharpoons HCO^- + H^+ + \overset{NH_3}{\downarrow} \\ \underset{\boxed{\text{Eggshell}}}{CO_3^{2-}} + \underset{\boxed{\text{Blood}}}{NH_4^+} + H^+ \quad (5)$$

This reaction is written on an equimolar basis with regard to ammonia input, but presumably larger inputs of this substance could assist in buffering the free protons produced by dissociation of carbonic acid. Simkiss (1975), however, has questioned whether ammonia plays a major buffering role on two grounds. Firstly, the quantity of the enzyme (adenosine deaminase) available for ammonia synthesis is low in the oviduct and must, therefore, be rate-limiting, especially considering the 100 mmol or so of protons produced when a chicken calcifies its eggshell. Secondly, if these ions were all titrated to ammonium, then there would not be any arteriovenous pH drop across the oviduct, rather the systemic acidosis would be manifested as the ammonium was metabolized by the liver. These changes do not follow the above pattern and we must conclude that ammonia plays, at most, a minor buffering role in the process of eggshell calcification. Equation (5) would satisfy the criteria for a derivation of carbonate from a non-plasma bicarbonate source; it would also provide a role for carbonic anhydrase and allow for the production of protons which could subsequently be secreted into the blood. Furthermore, it could provide a role for ammonia in precipitating carbonate, although this may not be particularly important.

Even the system outlined above might be an oversimplification of a complex process as there is now evidence that carbonic anhydrase exists as two iso-enzymes, at least in mammals, and possibly other tetrapods, an evolutionary modification which has been suggested as associated with sodium, potassium and chloride ion transport (Carter, 1972). It is likely that the activity of this enzyme in the shell gland is modulated by hormones and it has already been proposed that parathyroid hormone could exert an inhibitory influence on oviduct carbonic anhydrase to limit $CaCO_3$ secretion into the eggshell. In some mammals, e.g. rabbits, carbonic anhydrase in the uterine endometrium has been found to respond to sex hormones with oestrogens apparently inhibiting and progesterone markedly stimulating its activity (Carter, 1972). Similar responses in the avian shell gland as well as bone, if present, could have important functional significance in the hen's eggshell calcification cycle, and this would clearly be a fruitful area for further research.

Calcium Appetite in the Egg-laying Bird

In recent years, interest has developed in the possibility of a specific appetite for Ca in egg-laying hens. Hughes and Wood-Gush (1971) suggested that hens are able to vary their Ca intake in order to satisfy requirements of this electrolyte at particular times of Ca stress. The appetite appeared to be highly specific since there was no generalization to strontium, the element most closely associated with Ca, and was manifested whether Ca was in the form of a carbonate or lactate. It appeared to be a learned preference rather than an unlearned homeostatic control and was cue dependent, with visual and gustatory cues playing an important role in dietary selection.

In a later paper, Hughes (1972) demonstrated a circadian rhythmn of Ca intake apparently controlled in response to ovulation. Data from one of his experiments are shown in Fig. 59. Here the birds were placed in a Skinner box and trained to peck a key which, for each peck, gave five seconds' access to Ca grit, each peck (or reinforcement) denoting a unit of "calcium appetite". In birds which had previously ovulated, the Ca appetite was considerably increased during the last five hours or so of the light period; birds which had not ovulated showed no such increase. Since successive ovulations are staggered in chickens, the peak of dietary Ca intake occurred at a variable time in relation to ovulation, but always before the onset of eggshell calcification. From studies of the plasma Ca level in egg-laying birds (see above) we know that this would be at about its highest before the onset of eggshell calcification, hardly an appropriate stimulus for increased Ca intake. Alternatively, if we assume that a hypocalcaemic stimulus associated with the final stages of eggshell calcification triggers the appetite, this might be expected to occur earlier in the day than it does and not necessarily on the first ovulation day following a clutch pause. Again, the data in Fig. 59 do not fit such a picture.

Recently it was shown that infusion of bovine parathyroid hormone (1 U.S.P. U $kg^{-1} h^{-1}$) into chickens causes a significant decrease in dietary intake of a Ca supplement over a period of ten days compared with vehicle infused controls.

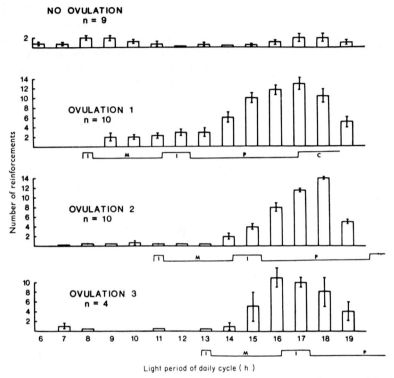

FIG. 59. Diurnal pattern of Ca intake in an individual hen under normal laying conditions. Columns represent the mean number of hourly reinforcements ($\pm$ s.e.). The estimated passage of the egg through the oviduct is shown below each histogram: Infundibulum (I), magnum (M), isthmus (I), plumping (P), calcification (C). Light period of daily cycle 0600 h–1800 h. (From Hughes, 1972.)

This effect was apparent in the absence of any noteworthy hypercalcaemic response to the hormone. Infusion of salmon calcitonin ($0 \cdot 2$ MRC U kg^{-1} h^{-1}) or of EDTA and injections of oestradiol diproprionate and/or testosterone proprionate did not significantly alter Ca intake (Joshua and Mueller, 1977). These results indicated that parathyroid hormone can affect a Ca appetite in some way, not necessarily by altering plasma Ca levels which, in turn, would regulate the Ca appetite. The intriguing possibility exists that parathyroid hormone might directly influence the central nervous system to induce a change in the pattern of behaviour. Perhaps the control of Ca appetite in hens is modulated by a variety of factors including both external (visual and gustatory) and internal (hormonal) stimuli. Endocrinologists are increasingly recognizing that hormones regulating systemic and metabolic parameters can also have neurotrophic influences resulting in behavioural patterns which may enhance endogenous homeostatic processes.

CONCLUSION

Clearly then, Ca metabolism in the egg-laying hen is a dynamic process which is influenced by a wide variety of factors both endogenous and exogenous. The improvements that can be made to eggshell quality are limited and any advances which are to be of commercial use are likely to be in the field of poultry husbandry, with respect to cage construction and suchlike. However, the study of a Ca metabolism in egg-laying birds is exciting and worthwhile in its own

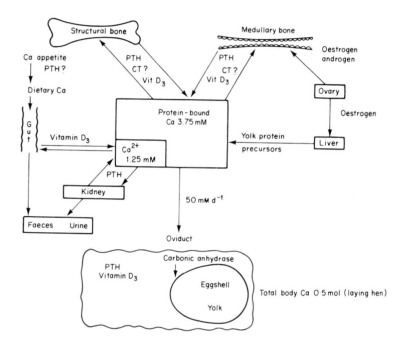

FIG. 60. Current concept of Ca metabolism in egg-laying hen with a total body Ca content of 0·5 mol. The major differences from mammalian species are: (1) the influence of gonadal hormones, (2) formation of medullary bone within marrow cavities of the long bones, (3) high turnover of Ca during egg production and (4) mixing of urine and faeces prior to excretion. High oestrogen secretion causes production of liver yolk protein precursors which bind Ca resulting in hypercalcaemia. Both oestrogens and androgens are required for medullary bone formation; under certain conditions (low oestrogen) medullary bone formation can be induced without accompanying hypercalcaemia. Medullary bone is assumed to serve as a store of labile Ca for use during eggshell formation. The latter is approx. 10% of total body store of Ca. Since Kenny and Dacke (1975), additional influences of parathyroid hormone (PTH) on oviduct and Ca appetite have been demonstrated.

182 CALCIUM REGULATION IN SUB-MAMMALIAN VERTEBRATES

right and investigations of the predictable movements of relatively large quantities of Ca and associated ions will lead us to a greater understanding of Ca metabolism in general. Dynamic considerations of Ca metabolism in egg-laying hens are summarized in Fig. 60 with possible points of hormone interaction shown.

14. Conclusions and Speculations

The aim of this book has been to discuss calcium (Ca) regulation in vertebrate classes other than mammals since these lower classes have been previously neglected in the literature. At the same time, reference is made to the mammalian system in order to clarify the sub-mammalian picture. Studies in non-mammals can be relevant to mammalian biology leading to a greater understanding of physiological processes in the latter class. An example of a clinical problem which might benefit from comparative studies is that of osteoporosis, a distressing syndrome affecting a large number of elderly women, which is undoubtedly related to changes in levels of circulating hormones following the menopause. Avian studies can provide useful models of this disease because the interrelation between gonadal and Ca regulating hormones in egg-laying hens is increasingly apparent. Hens become osteoporotic under certain conditions of Ca stress and it is likely that their hormonal balance plays a major contributory role in this situation.

Throughout the book the emphasis has been on collating data from several disciplines: physiologists can learn much from biochemists, pharmacologists, zoologists and palaeontologists, as well as contributing towards these disciplines. Obviously the ramifications of the subject discussed in this book are vast and new lines of possible research numerous. Three peripheral questions have dominated much of my thinking throughout the writing of the book and I believe them to be fundamental to the future understanding of the evolution of Ca regulation in vertebrates. They can be designated as (1) the phosphate problem, (2) the magnesium problem and (3) the acid-base problem. These will be considered below in relation to the central problem of (4) Ca regulation. The views presented in the following discussion are largely speculative.

The Phosphate Problem

In 1961, Poutard suggested that in early vertebrates bone may have arisen primarily as a store for the more basic insoluble phosphate ions rather than for Ca ions. This development would have given bony animals a definite advantage compared with non-bony animals, since environmental phosphorus shows a seasonal cycle of availability (see Chapters 1 and 2). Thus primitive vertebrates and most other animals which do not possess bone, in order to maintain a blood level of $1 \cdot 0$ mmol litre^{-1} inorganic phosphorus, must turn over enormous

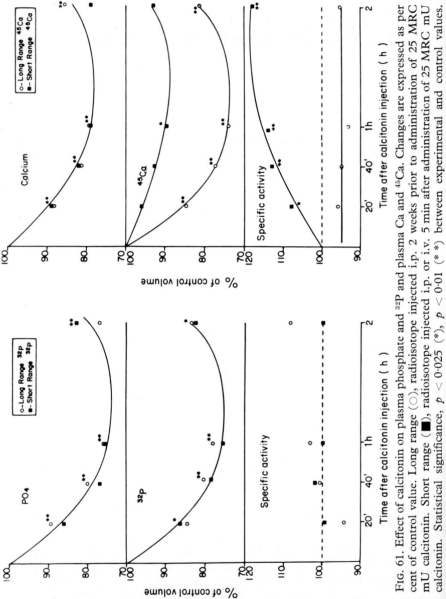

FIG. 61. Effect of calcitonin on plasma phosphate and ^{32}P and plasma Ca and ^{45}Ca. Changes are expressed as per cent of control value. Long range (○), radioisotope injected i.p. 2 weeks prior to administration of 25 MRC mU calcitonin. Short range (■), radioisotope injected i.p. or i.v. 5 min after administration of 25 MRC mU calcitonin. Statistical significance, $p < 0.025$ (*), $p < 0.01$ (**) between experimental and control values.

CONCLUSIONS AND SPECULATIONS   185

volumes of water and ingest plants or other animals continuously. However, higher vertebrates with bone are able to mobilize this tissue at times of low phosphorus availability to ensure a constant supply of the element. It is interesting to consider the turnover of phosphorus in mammalian bones, particularly with respect to the role of calcitonin. Talmage *et al.* (1972) examined the influence of salmon and porcine calcitonins on the disappearance of radiocalcium and radiophosphorus from the extracellular compartment in rats and their results are summarized in Fig. 61. Treatment of young rats with either type of calcitonin (25 MRC mU i.p.) caused a fall in the level of total plasma Ca and phosphate. If the rats were labelled with ^{45}Ca two weeks prior to calcitonin (long range) the hypocalcaemia was accompanied by a marked fall in plasma ^{45}Ca while the plasma ^{45}Ca/^{40}Ca specific activity remained constant. Simultaneous (short range) ^{45}Ca and calcitonin treatment resulted in a much smaller fall in plasma radiocalcium with a rise in specific activity. These results are consistent with an effect of the hormone inhibiting bone resorption.

The radiophosphorus results were entirely different. In both long and short term ^{32}P labelled rats a similar rapid fall in the plasma ^{32}P level occurred following treatment with the hormone. Plasma specific activity for both long and short term ^{32}P labelled rats, however, remained relatively constant (see Fig. 61). These results are consistent with a role for calcitonin stimulating exit of phosphate from the extracellular pool. A question remains as to the location of this phosphate disappearance, and while this is not yet clear, considerable evidence, cited by Talmage *et al.* (1972), indicates that at least a portion of the phosphate probably enters bone. More recent studies (Talmage, personal communication) support this concept.

Calcitonin probably evolved very early in vertebrate history; its presence in all extant classes other than Agnatha indicates its appearance at some stage between the Ostracodermi and Placodermi, or perhaps shortly after the first appearance of bone. Let us suppose that in these early vertebrates, phosphate levels in plasma were increased only spasmodically following feeding, while at other times the ion would be in rather short supply. It would, therefore, make sense to have a system in which during the infrequent periods of phosphate repletion, an animal could rapidly deposit the ion into a suitable store; then during longer periods of phosphate depletion, for instance, during nonfeeding phases, the phosphate store might be gradually leached out, thus maintaining a constant supply of this essential ion. Perhaps in teleost acellular bone we see the ultimate expression of phosphate storage over very long periods as suggested by Fleming (1967). Consistent with the idea that calcitonin is a modulator of phosphate storage into bony tissues is the fact that secretion of this hormone is triggered by secretagogues, such as gastrin, following feeding. Further evidence for an interaction between calcitonin and phosphate comes from mammalian studies. When the diet of rats is low in phosphate but adequate in Ca, hypocalcaemic responses to injected calcitonin are abolished; however, adequate phosphate, in combination with low Ca in the diet, does not inhibit the hormone response. Furthermore, addition of phosphate ions to the calcitonin injection medium greatly enhances the hypocalcaemic response in rats (Munson, 1976). This evidence indicates a primary action of calcitonin on phosphate metabolism, with Ca perhaps exhibiting a secondary response. Certainly it appears that in percentage terms the hypophosphataemic response to calcitonin is greater than the hypocalcaemic response, whatever the source or type of calcitonin used (Kennedy and Talmage, 1971). Diurnal changes in plasma phosphate levels in rats, rising during the fasting portion of the daily cycle and then falling abruptly during the feeding portion, support the hypothesis. These changes were reported to be less marked in rats which had been previously thyroidectomized, indicating a role for thyroid calcitonin in this cycle (Talmage et al., 1975).

While no feedback system has been reported with respect to the plasma phosphate level and calcitonin secretion, neither is the hypercalcaemic feedback mechanism for calcitonin particularly well established in sub-mammals such as teleost fish (see Chapter 10). Similarly, calcitonin secretion in birds is rather insensitive to changes in the plasma Ca concentration (see Chapter 13). It is becoming evident that the major modulators of calcitonin secretion are gut hormones released in the post-prandial situation (Talmage and Meyer, 1976). While this system would function reasonably well with respect to storage in bone of absorbed dietary Ca in tetrapods, particularly mammals, it is difficult to see why such a system would be needed in fish which have access to an abundant supply of environmental Ca readily available by direct exchange across the gills. The phosphate supply is another matter; this ion is only normally available in the diet as a more or less intermittent supply. It is appropriate to have a hormone system which rapidly stimulates removal of phosphate from the extracellular compartment to bone after its dietary absorption, since it might otherwise be quickly lost across the gills by diffusion.

Evidence for calcitonin having a major Ca regulating function, even in mammals, is far from convincing. Mammals are the only class in which the hormone has an obvious hypocalcaemic effect. It does have effects on bone in sub-mammalian species compatible with a reduction in mineral loss and osteolytic activity: in particular, the experiments of Lopez et al. (1971, 1976) indicate that in cellular boned teleosts, at least, the major role for calcitonin is coincident with an increase in bone apposition rather than with decreased resorption.

Could it be that calcitonin evolved primarily as a phosphate conserving hormone, facilitating the rapid storage of this ion into bone? Any effect on Ca might be secondary to this, so that only in mammals do we see any credible role for the hormone in Ca regulation, and even in this class the role may be subservient to that of vitamin D_3 and the parathyroids.

What of changes occurring in plasma calcitonin levels in bony fish, birds and mammals in relation to sexual activity? Salmon and Japanese quail hens exhibit pre-spawning surges in these levels associated with increased bone formation, while stags also show a plasma calcitonin surge at the onset of the rut (see Chapters 5, 10 and 13). Perhaps this surge is associated with increased storage of phosphate which can then be mobilized during spawning and reproduction when energy metabolism and the requirement for phosphorus must be at its highest.

It will be of paramount interest in the future to study the effects of calcitonin on phosphate rather than on Ca metabolism in sub-mammalian vertebrates, particularly in the three extant classes of fish.

Vitamin D_3 would function in the system outlined above since the active metabolite 1,25-dihydroxycholecalciferol not only stimulates gut Ca absorption but also phosphate absorption, at least in higher vertebrates (see Chapters 6 and 10). The recent demonstration by MacIntyre et al., (1976) of a hyperphosphataemic role for the metabolite in fish is particularly exciting in this respect as it suggests the phosphate absorption effect may predominate over any calcaemic influences of the vitamin in these species. Low levels of plasma phosphate also appear to be one of the stimuli for increased hydroxylation of 25- to 1,25-dihydroxycholecalciferol by the kidney. Presumably this helps prime the system to deal with an influx of dietary phosphate, at the same time increasing Ca influx and so raising the solubility product of these ions in extracellular fluid to facilitate their precipitation into bony tissues. Further studies into the role of vitamin D_3 and its metabolites in fish are desirable, particularly with regard to the effect of putative trophic factors such as Ca, phosphate, fish prolactin and even the newly discovered hypocalcin and hypercalcin, on the vitamin hydroxylation system.

The studies on the function of the pituitary in teleost fish could also fit into the scheme outlined above. Brehe and Fleming (1976) concluded that a pituitary hormone increases Ca uptake in the killifish (*Fundulus kansae*) in the summer and, to some extent, this uptake is reflected by the Ca (and presumably phosphorus) reservoirs of acellular bone, if not by the scales. Conversely, in the winter it seems the pituitary is involved in mobilization of mineral from acellular bone (see Chapter 10). While many authors are of the opinion that acellular fish bone is inert (see Chapter 4), it is becoming apparent that mineral can diffuse out of it, albeit at a very slow rate. Could acellular bone have evolved as a store for phosphorus for long-term needs, particularly to give a steady supply of the substance during winter months when environmental supplies may be short? Teleost scales, on the other hand, could serve as a much more labile mineral reserve for short-term regulation of Ca, phosphate and other ions. Acellular bone is specifically a teleost adaptation; perhaps this seemingly inert material is, in fact, a phosphorous store and as such has helped contribute towards the tremendous success of the teleost order.

The Magnesium Problem

It is more difficult to come to any firm conclusions concerning the relationship between magnesium and Ca metabolism. Previously in this book it was suggested that if early vertebrates had a problem repelling a high environmental Ca challenge, then surely they had a far greater one with the magnesium challenge. A controlled magnesium level is essential for the proper functioning of many enzyme systems as well as other biological processes. We see from an examination of magnesium levels in the extracellular fluids of animals that this divalent ion is regulated as closely as the Ca ion (Livingstone and Wacker, 1976), though how, and by what it is regulated, remains a puzzle.

The trouble with magnesium is that it is rather difficult to work with and until fifteen years ago, with the advent of atomic absorption spectrophotometry, there was no reliable method for accurately measuring magnesium in biological samples. Also, since radioactive isotopes of magnesium are not generally available, it is not possible to study exchanges of this ion between various pools. As yet, no hormonal system has been demonstrated as having a specific action concerned with the regulation of magnesium metabolism. Of the three major Ca regulating hormones in mammals, parathyroid hormone has the most obvious effect on magnesium, its general action and feedback control being rather similar to those for Ca (Vaughan, 1970). Calcitonin appears to have a slight hypomagnesaemic influence (Munson, 1976) but that of the vitamin D_3 system is not clear. Magnesium and Ca regulation are, to some extent, interdependent, so that Ca regulation is not maintained in the face of any marked abnormalities in the level of plasma magnesium (Vaughan, 1970). The disparate biochemical and physiological functions of the two ions, however, indicate a need for separate and specific regulatory systems.

A key study concerning magnesium homeostasis in sub-mammals is that of Oguro and Uchiyama (1975) on the role of the pituitary in electrolyte metabolism in Urodele amphibians. In this order, hypophysectomy leads to a marked decline in plasma magnesium as well as sodium and Ca levels (Chapter 11). If we extrapolate these results back to fish, it might be found that the pituitary has some hypermagnesaemic action, and if this is the case it is likely that the pituitary would play a role in magnesium homoestasis in euryhaline species during transfer from sea to fresh water. This certainly seems to be true for other ions such as Ca and sodium in fish (Bentley, 1976). Further studies are desirable with respect to the effect of specific pituitary hormones on magnesium homeostasis in fish and amphibia.

The Acid-base Problem

Next, the problem of acid-base balance, to which reference has been made in several chapters of this book. The most obvious question to be asked is, did parathyroid hormone first evolve in amphibia in an attempt by this class to control periods of respiratory acidosis with which they would have inevitably been faced? The notable work of Simkiss and of Robertson referred to in Chapter 11, strongly suggests a role for the massive $CaCO_3$ deposits of the anuran endolymphatic sacs in acid-base balance. These deposits could function as reserves of buffer base, thereby replacing reserves of bicarbonate normally found in the aquatic environment of fish. It is relevant to recall that the carbonate content of fish bone tends to be lower than that of tetrapods (see Chapter 4). The bicarbonate ion, as well as dissolved carbon dioxide, can be

CONCLUSIONS AND SPECULATIONS 189

exchanged across the fish gill epithelium (Maetz, 1974), and in fish, such as trout or carp, blood pH is high (7·37–8·00) while the PCO_2 (3·7–10·3 mm Hg) and bicarbonate (8·6–19·5 mmol litre^{-1}) are considerably lower than equivalent values for tetrapods (Neuman and Mulryan, 1968). Tetrapods can, therefore, be said to have a respiratory acidosis when compared with fish. As early amphibious vertebrates began to colonize land, they were faced with the problem of eliminating carbon dioxide leading to a build-up of protons and bicarbonate ions in their body fluids. It is interesting that lungfish (*Dipnoi*), which, like amphibia, breathe air for part of the time, also have comparatively well-developed endolymphatic sacs filled with calcareous material (see Chapter 4). We should remember the manner in which the carbonates are dissociated:

$$2CO_2 + 2H_2O \xrightleftharpoons[\text{carbonic anhydrase}]{} 2H_2CO_3 \rightleftharpoons 2H + 2HCO_3^-$$

$$\Updownarrow$$

$$2H + CO_3^{2-}$$

This shows the influence which a variation in pH of a buffered medium can have on the concentrations of ions involved in the chemical equilibria. The Henderson-Hasselbalch equation takes account of these variations:

$$pH = pK + \log \frac{(HCO_3^-)}{\alpha\, CO_2}$$

From this we can see that bone or endolymphatic carbonates form an important reserve of buffer base.

Parathyroid hormone appears to be the main hormone affecting turnover of endolymphatic lime sac deposits in anuran amphibians, although calcitonin and the vitamin D_3 metabolites may also have some influence. Devoid of their parathyroid glands, frogs are unable to compensate efficiently against an acidotic challenge (see Chapter 11).

It is quite possible that parathyroid hormone also mobilizes buffer-base reserves in higher vertebrates, but since these reserves exist mainly in bone, either as free carbonate or in the form of carbonate-hydroxyapatite, the responses are probably more covert than in species possessing discrete endolymphatic deposits. Wills (1970) suggested that not only bone carbonate, but also phosphate, represents an important reserve of buffer-base which can be mobilized by the hormone. The latter ion is a major intracellular, as well as urinary buffer, while the carbonate-bicarbonate system has a mainly extracellular function. As mentioned previously, carbonic anhydrase appears to play a role in the mediation of parathyroid hormone action in bone as well as other sites. The considerable evidence implicating carbonic anhydrase and carbon dioxide in parathyroid hormone mediated bone resorption has been reviewed by Kenny and Dacke (1975). It is likely that carbonic anhydrase is directly involved in the formation and resorption of bone $CaCO_3$.

In mammals, we see that the $CaCO_3$ fraction of bone is the first to be mobilized during metabolic acidosis. It would be of great interest if a differential effect of parathyroid hormone on the bone carbonate rather than phosphate fraction could be demonstrated: this might be accomplished by dual labelling of bone with $H^{32}PO_4$ and $^{14}CO_3$ with a subsequent analysis of mobilization of the isotopes by the hormone. Certainly in acidotic frogs which mobilize Ca, loss of phosphate is not varied, suggesting a carbonate rather than hydroxyapatite source of buffer which, in these species, would be represented by the endolymphatic lime sacs (see Chapter 11).

While the parathyroids have an effect on amphibian endolymphatic sacs or bone compatible with a role in counteracting acidosis, the effects of the hormone on kidney, at least in mammals, are ambiguous. Thus in dogs and humans, injection of 100 U.S.P. U of pure bovine parathyroid hormone leads to an increase in urine pH, bicarbonate and phosphate, while titrateable acid and ammonium are decreased (Hellman et al., 1965). The effects on acid excretion are rapid, transient and apparently separate from those on phosphate. The former effects are probably mediated by an inhibition of renal carbonic anhydrase (see Chapter 6). Whereas the effect on phosphate excretion could represent an increase in urine buffering capacity, those on urine pH, bicarbonate and total acid are not consistent with a role for the hormone in counteracting acidosis, since these should involve conservation of bicarbonate coupled with increased acid excretion (Pitts, 1968). It remains to be seen whether or not a more chronic infusion of parathyroid hormone would cause the same renal responses with respect to acid excretion. Perhaps the transient responses seen during acute parathyroid treatment partly reflect an increase in blood buffering capacity mediated by the action of the hormone on calcified tissues. By reference to the above equations, it can be argued that the increased excretion of bicarbonate in urine allows carbonate reserves to be mobilized from calcified tissues and these will act as a more powerful buffer than the plasma bicarbonate which, itself, is a form of buffered acid.

Another paradox is that of hyperparathyroid disease in humans, in which situation it is well known that an accompanying metabolic acidosis occurs. This appears to result from the effect of parathyroid hormone on the renal tubule (Cuisinier-Gleizes et al., 1975) and is presumably mediated by the inhibition of renal carbonic anhydrase. It should be stressed, however, that this is a pathological response to abnormally high circulating levels of parathyroid hormone and does not necessarily occur with more normal levels of the hormone.

It is not clear whether the renal carbonic anhydrase response discussed above simply reflects a generalized action of this enzyme to parathyroid hormone with little functional significance in the mammalian kidney, but more relevance in tissues such as bone, endolymphatic epithelium and avian oviduct. In the higher vertebrates, at least, kidney is a major organ for acid-base regulation and would, therefore, be expected to show rather different responses to parathyroid hormone as a regulator of acidosis.

Frazier (1976) studied the responses of an isolated toad bladder preparation

CONCLUSIONS AND SPECULATIONS 191

to partially purified, bovine parathyroid hormone (see Chapter 11) and obtained results opposite from those found with mammalian kidney in that the hormone actually stimulated secretion of protons by the bladders. The response was elicited with doses as low as 260 U.S.P. mU ml^{-1} in the solution bathing the serosal surface and appeared to involve the action of 3′,5′-cyclic AMP. This response is much more in keeping with the concept of parathyroid hormone controlling acidosis. Frazier considers the toad bladder to have a function analogous to that of the distal tubule of the mammalian nephron, while the inhibitory effect of the hormone on bicarbonate absorption in mammalian kidney discussed above is thought to be localized at the proximal tubule (Frazier, 1976).

When the secretion of parathyroid hormone is considered, at least in mammals (see Chapter 5), we again find a paradox. In one investigation it was suggested that secretion of this hormone in dogs could be stimulated during acute metabolic alkalosis and inhibited during metabolic acidosis (Fujita *et al.*, 1965). This finding is hardly compatible with the concept of a negative feedback regulation of the hormone secretion during acidosis although it might be due to a change in pH affecting ionized Ca, or through an action on buffer phosphate which, in turn, would affect Ca levels (see Chapter 5).

In a more recent study involving human volunteers subjected to chronic metabolic (ammonium chloride) acidosis, indirect evidence indicated an increase in parathyroid activity since the experimental subjects apparently entered a phase of negative Ca balance, while urinary phosphate and hydroxyproline excretion were increased. Plasma Ca levels were not markedly changed during the acidosis but plasma levels of immunoreactive parathyroid hormone were measured in four of the five subjects and showed a tendency to increase during the acidosis (Wachman and Bernstein, 1970). Unfortunately, no proper control experiments were carried out in this study and the inappropriate statistical analyses detract from what could be an interesting line of investigation.

A similar series of experiments in rats under more controlled conditions gave comparable results (Cuisinier-Gleizes *et al.*, 1975). These workers fed rats ammonium chloride for seven weeks to promote a metabolic acidosis. The rats efficiently compensated for this by increased urine levels of Ca, phosphate, hydroxyproline and total acid excretion (see Tables XXXIII and XXXIV). If the rats were either parathyroidectomized or thyroparathyroidectomized prior to induction of acidosis, they lost their ability to counteract this challenge as seen from the fall in serum pH and carbon dioxide (Table XXXIII). Changes in serum and urine parameters (Tables XXXIII and XXXIV) indicate that bone resorption was more marked in the parathyroidectomized and thyroparathyroidectomized rats than controls during acidosis. These findings led the authors to propose that the parathyroids regulate a fall in serum pH by a mechanism which does not involve general mobilization of bone mineral, but rather the mobilization of carbonate-bicarbonate buffer from the bone, which, in turn, helps counteract the acidosis.

TABLE XXXIII

Effect of parathyroidectomy and thyroparathyroidectomy on serum Ca and acid-base, regulatory parameters in rats

	Ca (mmol litre^{-1})	Pi (mmol litre^{-1})	Alkaline Phosphatase (i.u. litre^{-1})	pH	CO_2 (mmol litre^{-1})
PTX					
NH$_4$Cl	1·88 ± 0·08	0·60 ± 0·02	433 ± 39	7·23 ± 0·03	10·8 ± 0·9
Controls	1·55 ± 0·13	0·65 ± 0·05	386 ± 61	7·39 ± 0·02	17·6 ± 1·2
p	<0·05	>0·05	>0·05	<0·01	<0·001
TPTX					
NH$_4$Cl	2·40 ± 0·07	0·61 ± 0·03	296 ± 26	7·12 ± 0·01	—
Controls	1·73 ± 0·12	0·68 ± 0·07	245 ± 31	7·34 ± 0·03	—
p	<0·01	>0·05	>0·05	<0·001	
Sham-operated					
NH$_4$Cl	2·48 ± 0·02	0·47 ± 0·01	371 ± 48	7·40 ± 0·01	14·0 ± 1·1
Controls	2·48 ± 0·03	0·44 ± 0·03	332 ± 72	7·42 ± 0·01	18·0 ± 1·3
p	>0·05	>0·05	>0·05	>0·05	<0·05

All values mean ± s.e., PTX = parathyroidectomy, TPTX = thyroparathyroidectomy.
(Modified after Cuisinier-Gleizes *et al.*, 1975.)

TABLE XXXIV

Effect of parathyroidectomy and thyroparathyroidectomy on urine Ca and acid-base regulatory parameters in rats

	Ca (mmol 24h^{-1})	Pi (mmol 24h^{-1})	(OH)-Proline (mg 24h^{-1})	Titrateable acid (mmol 24h^{-1})	NH$_3$ (mmol 24h^{-1})
PTX					
NH$_4$Cl	0·73 ± 0·17	0·50 ± 0·05	0·842 ± 0·090	0·274 ± 0·030	4·64 ± 0·46
Controls	0·05 ± 0·01	0·28 ± 0·03	0·368 ± 0·045	Trace	0·57 ± 0·07
p	<0·01	<0·01	<0·001		<0·001
TPTX					
NH$_4$Cl	1·15 ± 0·34	0·72 ± 0·10	0·968 ± 0·102	0·423 ± 0·070	4·56 ± 0·54
Controls	0·05 ± 0·01	0·32 ± 0·04	0·409 ± 0·035	0·016 ± 0·001	0·30 ± 0·03
p	<0·01	<0·01	<0·001	<0·001	<0·001
Sham-operated					
NH$_4$Cl	0·93 ± 0·05	0·74 ± 0·03	0·702 ± 0·041	0·563 ± 0·023	5·32 ± 0·24
Controls	0·08 ± 0·01	0·53 ± 0·03	0·504 ± 0·014	0·025 ± 0·022	0·50 ± 0·04
p	<0·001	<0·001	<0·001	<0·001	<0·001

All values mean ± s.e., PTX = parathyroidectomy, TPTX = thyroparathyroidectomy.
(Modified after Cuisinier-Gleizes *et al.*, 1975.)

A key feature of the experimental data discussed above is that they all concern responses to acidosis or alkalosis of metabolic rather than respiratory origin. The hypothesis presented in this chapter is that parathyroid hormone evolved in amphibia in response to acidotic challenge of respiratory origin since the transition from water- to airbreathing would have led to problems in disposal of waste carbon dioxide. It is to be hoped that in the future, parathyroid secretion will be reinvestigated during both acute and chronic changes in acid-base status induced by both metabolic and respiratory changes. An important distinction between metabolic and respiratory acidosis and alkalosis lies in the levels of bicarbonate and PCO_2 in the blood and it is possible that changes in these parameters themselves interfere with total and ionized levels of Ca in the plasma.

Future development of suitable parathyroid hormone assays will allow similar investigations to those outlined above in a variety of tetrapods including not only mammals, but lower classes, particularly amphibians. In mammals, hypocalcaemic feedback is undoubtedly the major controlling factor in parathyroid hormone secretion. Whether or not this is so in lower tetrapods, remains to be seen. It is possible that the role of parathyroid hormone in acid-base balance and Ca metabolism has radically altered between the evolution of amphibia and more advanced tetrapods. Another possibility, suggested by Robertson (personal communication), is that this hormone is involved in counteracting a metabolic acidosis in amphibia and probably higher tetrapods, which occurs during hibernation. Hence as carbohydrate stores become depleted, the animals could become ketotic and, as a result, acidotic. Mobilization of endolymphatic or bone buffer base reserve would help counteract this acidosis. It is not yet clear whether the spring cycle of regeneration of parathyroid glands (see Chapters 5 and 11) would follow such a period of acidosis, or run concurrently with it.

Acid-base regulation in mammals is sophisticated with the lungs and kidneys playing a major role: in amphibia, these organs are more primitive and probably less important in controlling pH. More work is needed on the influence of the parathyroids in acid-base balance in all classes of tetrapods before we can reach any consensus about the primary metabolic function of this gland.

The Calcium Problem

In this final chapter it has been suggested that two of the major hormones supposedly involved in vertebrate Ca homeostasis, i.e. calcitonin and parathyroid hormone, may, in reality, have arisen in answer to the requirements of phosphate and acid-base metabolism respectively. It does not follow, however, that these hormones are not intimately involved in Ca metabolism, and, with respect to parathyroid hormone in mammals, at least, the fluctuations in plasma Ca levels which would occur in the absence of the parathyroid glands would be deleterious to life.

CONCLUSIONS AND SPECULATIONS 195

Can we identify any hormone which has evolved primarily as a regulator of Ca metabolism? The answer is probably "yes". The most important and specific of the Ca regulating hormones appears to be the active vitamin D_3 metabolite, 1,25-dihydroxycholecalciferol. It is not yet clear how far back in vertebrate phylogeny, evolution of the enzymic machinery necessary for the two hydroxylation steps in vitamin D_3 metabolism occurred. They are clearly present in the Osteichthyes but further work is needed to establish their presence or absence in the Chondrichthyes and, perhaps, even the Agnatha. Evidence has been presented (Chapter 5) which indicates that this final hydroxylation step is regulated by a variety of trophic factors including parathyroid hormone, prolactin, gonadal hormones and inorganic phosphate.

Bone is not a requisite adjunct for the hypercalcaemic function of 1,25-dihydroxycholecalciferol which might, therefore, be active in the extant non-bony lower classes of vertebrates. From the limited evidence available, however, it would seem that the vitamin D system achieves greater importance in tetrapods than in fish. Perhaps this is related to the fact that tetrapods have a much more intermittent supply of Ca than do fish. It is not surprising that the most important target organ for the active vitamin D_3 metabolite is the gut, although the fact that bone and kidney also respond to this substance suggests that it does not merely function to increase the efficiency of dietary Ca absorption in the post-prandial situation.

In more primitive vertebrates other hormones may play a more important part in regulation of the Ca ion; thus the role of the pituitary in Ca regulation is being increasingly recognized.

While prolactin has, for some time, been known to have hypercalcaemic effects in the vertebrates, this response could reflect more generalized actions of the hormone on water and electrolyte metabolism. The recent discovery that prolactin has a trophic action on vitamin D_3 metabolism (see Chapter 5) should lead to renewed interest in its hypercalcaemic actions, particularly since interaction of Ca metabolism with sexual status has many implications in mammals as well as sub-mammalian species.

The recent work by Parsons and his associates suggesting that the pituitary contains a parathyroid-like substance, hypercalcin (see Chapter 5), must spark off further research in this gland. Is hypercalcin the archetypal Ca regulating hormone? If so, how far down the vertebrate hierarchy or even beyond, will it be found?

Studies on the fish Ca regulating hormone, hypocalcin, should also increase now that preliminary attempts to extract and assay the active substance have met with success. Although this hormone is apparently only present in the Osteichthyes, it would be of interest to test its pharmacological properties in other vertebrate classes.

There is no doubt that even the most primitive extant vertebrates, the Cyclostomes, are able actively to regulate Ca ions in their extracellular fluids, a factor which enables many of them to lead a euryhaline mode of life: surely an advanced development in evolutionary terms. They are able to accomplish this in the apparent absence of the three hormones which are considered to regulate Ca in higher vertebrates, i.e. parathyroid hormone, calcitonin and the vitamin D_3 metabolites; and also in the absence of bone, the tissue which many workers consider to be the primary target organ for these "calcium regulating" hormones. Are the lampreys and hagfish trying to tell us something?

In a book of this size, coverage of the field is, of necessity, fairly superficial. The fascinating subject of Ca metabolism in vertebrate embryonic or larval forms has not been covered since this specialized field was reviewed in depth by Simkiss (1967). The main advances since then have been with respect to the avian embryo. No attempt has been made to consider abnormal Ca metabolism, for instance with regard to the effect of toxic pollutants such as the organo-chlorine insecticides. This is an area which, by itself, now warrants a review. It is hoped, however, that the present book provides some background information to workers in the aforementioned fields.

References

Atkins, D. (1976). *J. Endocr.* **69**, 28p.
Aurbach, G. D. and Chase, L. R. (1976). *In* "Handbook of Physiology." (R. O. Greep, E. B. Astwood and G. D. Aurbach, Eds) Vol. 7, pp. 353–382. Williams and Wilkins, Baltimore.
Aurbach, G. D., Keutmann, H. T., Niall, H. D., Tregear, G. W., O'Riordan, J. L. H., Marcus, R., Marx, S. J. and Potts, J. T. (1972). *Rec. Prog. Hormone Res.* **28**, 353–392.
Bailey, R. E. (1957). *J. exp. Zool.* **136**, 455–469.
Baksi, S. N. and Kenny, A. D. (1976a). *Fedn Proc.* **35**, 662.
Baksi, S. N. and Kenny, A. D. (1976b). Programme and Abstracts of 58th Annual Meeting U.S. Endocrine Society, p. 262.
Baksi, S. N. and Kenny, A. D. (1976c). *Pharmacologist* **18**, 661.
Baksi, S. N. and Kenny, A. D. (1977). *Pharmacologist* **19**, 335.
Ball, J. N. and Ingleton, P. M. (1973). *Gen. comp. Endocr.* **20**, 312–325.
Bannister, D. W. and Candlish, J. K. (1973). *Br. Poult. Sci.* **14**, 121–125.
Bar, A. and Hurwitz, S. (1973). *Comp. biochem. Physiol.* **45A**, 579–586.
Bar, A., Dubrov, D., Eisner, U. and Hurwitz, S. (1976). *Poult. Sci.* **55**, 622–628.
Barker, W. C. and Dayhof, M. O. (1972). *In* "Atlas of Protein Sequence and Structure." (M. O. Dayhof, Ed.) pp. 101–110. National Biomedical Research Foundation, Washington, D.C.
Bates, R. F. L., Bruce, J. and Care, A. D. (1969). *J. Endocr.* **45**, XIV–XV.
Bates, R. F. L., Care, A. D., Peacock, M., Mawer, E. B. and Taylor, C. M. (1975). *J. Endocr.* **64**, 6p.
Beck, N., Kim, K. S., Wolak, M. and Davis, B. B. (1974). *Clin. Res.* **22**, 515a.
Bélanger, L. F. and Copp, D. H. (1972). *In* "Calcium, Parathyroid Hormone and the Calcitonins." Proceedings of 4th Parathyroid Conference, Chapel Hill, 1971. (R. V. Talmage and P. L. Munson, Eds) pp. 41–50. Excerpta Medica, Amsterdam.
Bélanger, L. F. and Drouin, P. (1966). *Can. J. physiol. Pharmacol.* **44**, 919–922.
Bélanger, L. F., Dimond, M. T. and Copp, D. H. (1973). *Gen. comp. Endocr.* **20**, 297–304.
Bell, N. H. (1970). *J. clin. Invest.* **49**, 1368–1373.
Bentley, P. J. (1976). "Comparative Vertebrate Endocrinology." Cambridge University Press, Cambridge, London, New York and Melbourne.
Berg, A. (1968). *Mem. Ist. ital. Idrobiol.* **23**, 161–196.
Bernstein, D. S. and Handler, P. (1958). *Proc. Soc. exp. Biol. Med.* **99**, 339–340.
Bianchi, C. P. (1968). "Cell Calcium." Butterworth, London.
Bijvoet, O. L. M., Van Der Sluys Veer, J., de Vries, H. and Van Koppen, A. T. H. (1971). *New Engl. J. Med.* **284**, 681–687.
Blitz, R. M. and Pellegrino, E. D. (1961). *J. Bone J. Surg.* (Amer.) **51**, 456–466.
Blondin, G. A., Kulkarni, B. D. and Nes, W. R. (1967). *Comp. biochem. Physiol.* **20**, 379–390.

Bloom, M. A., Bloom, W., Domm, L. V. and McLean, F. C. (1940). *Anat. Rec.* **78**, 143.
Bloom, M. A., McLean, F. C. and Bloom, W. (1942). *Anat. Rec.* **83**, 99–120.
Bloom, W., Bloom, M. A. and McLean, F. C. (1941). *Anat. Rec.* **81**, 443–75.
Blum, J. W. Mayer, G. P. and Potts, J. T. (1974). *Endocrinology* **95**, 84–92.
Boelkins, J. N. and Kenny, A. D. (1973). *Endocrinology* **92**, 1754–1760.
Bormancin, M., Cuthbert, A. W. and Maetz, J. (1972). *J. Physiol.* **222**, 487–496.
Boschwitz, D. and Bern, H. A. (1971). *Gen. comp. Endocr.* **17**, 586–588.
Brand, R. S. and Raisz, L. G. (1972). *Endocrinology* **90**, 479–487.
Brehe, J. E. and Fleming, W. R. (1976). *J. comp. Physiol.* **110**, 159–169.
Brown, D. M., Perey, D. Y. E., Dent, P. B. and Good, R. A. (1969). *Proc. Soc. exp. biol. Med.* **130**, 1001–1004.
Brown, D. M., Perey, D. Y. E. and Jowsey, J. (1970). *Endocrinology* **87**, 1282–1291.
Brumbaugh, P. F. Haussler, D. H., Bressler, R. and Haussler, M. R. (1974a). *Science* **183**, 1089–1091.
Brumbaugh, P. F., Haussler, D. H., Bursac, K. M. and Haussler, M. R. (1974b). *Biochemistry* **13**, 4091–4097.
Buchanan, G. D. (1961). *In* "The Parathyroids." (R. O. Greep and R. V. Talmage, Eds) pp. 334–352. Thomas, Springfield, Illinois.
Budde, M. L. (1958). *Growth* **22**, 73–92.
Butler, D. G. (1969). *J. Fish Res. Bd Can.* **26**, 639–654.
Calamy, H. and Barlet, J. P. (1970). *C. r. Acad. Sci.* **271**, Series D, 2153–2156.
Campbell, J. W. and Speeg, K. V. (1969). *Nature* **224**, 222–236.
Candlish, J. K. (1970). *Comp. biochem. Physiol.* **32**, 703–707.
Candlish, J. K. and Taylor, T. G. (1970). *J. Endocr.* **48**, 143–144.
Care, A. D. and Bates, R. F. L. (1972). *Gen. comp. Endocr.* Suppl. **3**, 448–458.
Care, A. D., Cooper, C. W., Duncan, T. and Orimo, H. (1968). *Endocrinology* **83**, 161–169.
Care, A. D., Bates, R. F. L. and Gitelman, H. J. (1969). *J. Endocr.* **43**, lv–lvi.
Care, A. D., Bates, R. F. L. and Gitelman, J. J. (1970). *J. Endocr.* **48**, 1–15.
Care, A. D., Bruce, J. B., Boelkins, J. N., Kenny, A. D., Conaway, H. H. and Anast, C. S. (1971). *Endocrinology* **89**, 262–271.
Care, A. D., Bates, R. F. L., Swaminathan, R., Scanes, C. G., Peacock, M., Mawer, E. B., Taylor, C. M., De Luca, H. F., Tomlinson, S. and O'Riordan, J. L. H. (1975). *In* "Calcium Regulating Hormones." Proceedings of the 5th Parathyroid Conference. (R. V. Talmage, M. Owen and J. A. Parsons, Eds) pp. 100–110. Excerpta Medica, Amsterdam.
Carlstrom, D. (1963). *Biol. Bull. Woods Hole* **125**, 441–463.
Carter, M. J. (1972). *Biol. Rev.* **47**, 465–513.
Carter, T. C. (1971). *Br. Poult. Sci.* **12**, 379–385.
Chan, D. K. O. (1969). *In* "Progress in Endocrinology." Proceedings of 3rd International Congress Endocrinology, Mexico, 1968. (C. Gual and F. J. G. Ebling, Eds) pp. 709–716. Excerpta Medica, Amsterdam and New York.
Chan, D. K. O. (1972). *Gen. comp. Endocr.* Suppl. **3**, 411–420.
Chan, D. K. O., Chester-Jones, I., Henderson, I. W. and Rankin, J. C. (1967). *J. Endocr.* **37**, 297–317.
Chan, D. K. O., Chester-Jones, I. and Smith, R. N. (1968). *Gen. comp. Endocr.* **11**, 243–245.
Chartier, M. M. (1973). *C. r. Acad. Sci.* **276**, 785–788.
Chartier, M. M., Milet, C., Lopez, E., Lallier, F., Martelly, E. and Warrat, S. (1977). *J. Physiol.* Paris **73**, 23–36.

Chase, L. R. and Aurbach, G. D. (1968). *In* "Parathyroid Hormone and Thyrocalcitonin." Proceedings of 3rd Parathyroid Conference, Montreal, 1967. (R. V. Talmage and L. F. Bélanger, Eds) pp. 247–257. Excerpta Medica, Amsterdam and New York.
Chase, L. R. and Obert, K. A. (1975). *Metabolism* **24,** 1067–1071.
Chrichton, M. I. (1935). *Salm. Fish.* Edinburgh **4,** 1–8.
Citron, L., Exley, D. and Hallpike, C. S. (1956). *Brit. Med. Bull.* **12,** 101–104.
Clark, I. and Rivera-Cordero, F. (1973). *Endocrinology* **92,** 62–71.
Clark, N. B. (1965). *Gen. comp. Endocr.* **5,** 297–312.
Clark, N. B. (1971). *J. exp. Zool.* **178,** 9–14.
Clark, N. B. (1972). *Gen. comp. Endocr.* Suppl. **3,** 430–440.
Clark, N. B. and Fleming, W. R. (1963). *Gen. comp. Endocr.* **3,** 461–467.
Clark, N. B., Pang, P. K. T. and Dix, M. W. (1969). *Gen. comp. Endocr.* **12,** 614–618.
Coates, M. E. (1971). *In* "Physiology and Biochemistry of the Domestic Fowl." (D. J. Bell and B. M. Freeman, Eds) Vol. 1, pp. 373–396. Academic Press, London and New York.
Cohn, D. V., MacGregor, R. R., Chu, L. L. H., Kimmel J. R. and Hamilton, J. W. (1972). *Proc. natn. Acad. Sci.* Washington **69,** 1521–1525.
Collip, J. B. (1925). *J. biol. Chem.* **63,** 395–438.
Comar, C. L. and Driggers, J. C. (1949). *Science* **109,** 282.
Cooper, C. W., Schwesinger, W. H., Mahgoub, A. M. and Ontjes, D. A. (1971). *Science* **172,** 1238–1240.
Copp, D. H. (1968). *In* "Parathyroid Hormone and Thyrocalcitonin (Calcitonin)." Proceedings of 3rd Parathyroid Conference, Montreal, 1967. (R. V. Talmage and L. F. Bélanger, Eds) pp. 25–39. Excerpta Medica, Amsterdam and New York.
Copp, D. H. (1969a). *In* "Fish Physiology." (W. S. Hoar and D. J. Randall, Eds) Vol. II, pp. 377–398. Academic Press, London and New York.
Copp, D. H. (1969b). *J. Endocr.* **43,** 137–161.
Copp, D. H. (1972). *Gen. comp. Endocr.* Suppl. **3,** 441–447.
Copp, D. H. (1976). *In* "Handbook of Physiology." (R. O. Greep, E. B. Astwood and G. D. Aurbach, Eds) Vol. 7, pp. 431–442. Williams and Wilkins, Baltimore.
Copp, D. H., Cameron, E. C., Cheney, B. A., Davidson, A. G. F. and Henze, K. G. (1962). *Endocrinology* **70,** 638–649.
Copp, D. H., Cockcroft, D. W. and Kueh, Y. (1967a). *Can. J. physiol. Pharmacol.* **45,** 1095–1099.
Copp, D. H., Cockcroft, D. W. and Kueh, Y. (1967b). *Science* **158,** 924–926.
Copp, D. H., Cockcroft, R. W., Kueh, Y. and Melville, M. (1968). *In* "Calcitonin, Proceedings Symposium Thyrocalcitonin and the C Cells, 1967." (S. Taylor and R. Frazer, Eds) pp. 306–321. Heinemann, London.
Copp, D. H., Brooks, C. E., Low, B. S., Newsome, F., O'Dor, R. K., Parkes, C. O., Walker, V. and Watts, E. G. (1970). *In* "Calcitonin 1969," Proceedings of 2nd Symposium. (S. Taylor and R. Frazer, Eds) pp. 281–294. Heinemann, London.
Copp, D. H., Bélanger, L. F., Dimond, M., Newsome, F., Ng, D. and O'Dor, R. K. (1972). *In* "Endocrinology 1971." Proceedings of 3rd International Symposium, London. (S. Taylor, Ed.) pp. 29–38. Heinemann, London.
Corradino, R. A., Wasserman, R. H., Pubals, M. H. and Chang, S. I. (1968). *Arch. biochem. Biophys.* **125,** 378–380.
Cortelyou, J. R. (1960). *Anat. Rec.* **137,** 346.
Cortelyou, J. R. (1962a). *Endocrinology* **70,** 618–621.
Cortelyou, J. R. (1962b). *Amer. Zool.* **2,** 400.

Cortelyou, J. R. (1967). *Gen. comp. Endocr.* **9,** 234-240.
Cortelyou, J. R. and McWhinnie, D. J. (1967). *Amer. Zool.* **7,** 843-855.
Cuisinier-Gleizes, P., George, A., Thomasset, M. and Mathieu, H. (1975). *C. r. Acad. Sci. Paris* t. **280,** Series D, 2145-2148.
Cutler, G. B., Habener, J. F., Dee, P. C. and Potts, J. T. (1974). *FEBS Letts* **38/2,** 209.
Dacke, C. G. (1970). PhD. Thesis, University of Reading.
Dacke, C. G. (1972). *Proc. Trans. Missouri. Acad. Sci.* **6,** 178.
Dacke, C. G. (1975). *J. Physiol.* **246,** 75-76p.
Dacke, C. G. (1976). *J. Endocr.* **71,** 239-243.
Dacke, C. G. (1977). *J. Physiol* **273,** 4-5p.
Dacke, C. G. and Kenny, A. D. (1971). *Fedn Proc.* **30,** 417.
Dacke, C. G. and Kenny, A. D. (1973). *Endocrinology* **92,** 463-470.
Dacke, C. G., Fleming, W. R. and Kenny, A. D. (1971). *Physiologist* **14,** 127.
Dacke, C. G., Boelkins, J. N., Smith, W. K. and Kenny, A. D. (1972). *J. Endocr.* **54,** 369-370.
Dacke, C. G., Musacchia, X. J., Volkert, W. A. and Kenny, A. D. (1973). *Comp. biochem. Physiol.* **44A,** 1267-1275.
Dacke, C. G., Furr, B. J. A., Boelkins, J. N. and Kenny, A. D. (1976). *Comp. biochem. Physiol.* **55A,** 341-344.
Deftos, L. J., Murray, T. M., Powell, D., Habener, J. F., Singer, F. R., Mager, G. P. and Potts, J. T. (1972a). In "Calcium, Parathyroid Hormone and the Calcitonins." Proceedings of IV Parathyroid Symposium, Chapel Hill, 1972. (R. V. Talmage and P. L. Munson, Eds) pp. 140-151. Excerpta Medica, Amsterdam and New York.
Deftos, L. J., Watts, E. G. and Copp, D. H. (1972b). In "IVth International Congress on Endocrinology." International Congress Series **256,** 180. Excerpta Medica, Amsterdam and New York.
Deftos, L. J., Watts, E. G., Copp, D. H. and Potts, J. T. (1974). *Endocrinology* **94,** 155-160.
De Luca, H. F. (1976). In "Handbook of Physiology." (R. O. Greep, E. B. Astwood and G. D. Aurbach, Eds) Vol. 7, pp. 265-280. Williams and Wilkins, Baltimore.
Dempster, W. T. (1930). *J. Morph.* **50,** 7-126.
Dent, P. B., Brown, D. M. and Good, R. A. (1969). *Endocrinology* **85,** 582-585.
Dessauer, H. C. and Fox, W. (1959). *Am. J. Physiol.* **197,** 360-366.
Deville, J. and Lopez, E. (1970). *C. r. Acad. Sci. Paris* **270,** 2347-2350.
Dougherty, W. (1973). *Z. Zellforsch.* **146,** 167-175.
Dudley, J. (1942). *Am. J. Anat.* **71,** 65-97.
Dufresne, L. R. and Gitelman, H. J. (1972). In "Calcium, Parathyroid Hormone and the Calcitonins." Proceedings of 4th Parathyroid Conference, Chapel Hill, 1971. (R. V. Talmage and P. L. Munson, Eds) pp. 202-206. Excerpta Medica, Amsterdam and New York.
Dugal, L. P. (1939). *J. cell. Comp. Physiol.* **13,** 235-251.
Edgren, R. A. (1960). *Comp. biochem. Physiol.* **1,** 213-217.
Erulkar, S. D. and Maren, T. H. (1961). *Nature* **189,** 459-460.
Feinblatt, J. D. and Raisz, L. G. (1972). In "Calcium, Parathyroid Hormone and the Calcitonins." Proceedings of 4th Parathyroid Conference, Chapel Hill, 1971. (R. V. Talmage and P. L. Munson, Eds) p. 51. Excerpta Medica, Amsterdam and New York.
Fenwick, J. C. (1974). *Gen. comp. Endocr.* **23,** 127-135.
Fenwick, J. C. (1975). *Gen. comp. Endocr.* **25,** 60-63.
Fenwick, J. C. (1976). *Gen. comp. Endocr.* **29,** 383-387.

Fenwick, J. C. and Forster, M. E. (1972). *Gen. comp. Endocr.* **19**, 184–191.
Fenwick, J. C. and So, Y. P. (1974). *J. exp. Zool.* **188**, 125–131.
Fleisch, H. (1964). *Clin. Orthop.* **32**, 170–180.
Fleming, W. R. (1967). *Am. Zool.* **7**, 835–842.
Fleming, W. R. and Meier, A. H. (1961). *Comp. biochem. Physiol.* **2**, 1–7.
Fleming, W. R., Stanely, J. G. and Meier, A. H. (1964). *Gen. comp. Endocr.* **4**, 61–67.
Fleming, W. R., Brehe, J. and Hanson, R. (1973). *Am. Zool.* **13**, 793–797.
Fontaine, M. (1956). *Mem. Soc. Endocrinol.* **5**, 69–81.
Fontaine, M. (1964). *C. r. hebd. Séanc. Acad. Sci.* Paris **259**, 875–878.
Fontaine, M. (1967). *C. r. hebd. Séanc. Acad. Sci.* Paris **264**, 736–737.
Fontaine, M., Chartier-Baraduc, M., Deville, J., Lopez, E. and Poncet, M. (1969). *C. r. hebd. Séanc. Acad. Sci.* Paris **268**, 1958–1961.
Fontaine, M., Delerue, N., Martelly, E., Marchelidon, J. and Milet, C. (1972). *C. r. Acad. Sci.* **275**, Series D, 1523–1528.
Fontaine, M., Deville, J. and Lopez, E. (1973). Oceanography of the South Pacific 1972, comp. R. Frazer, New Zealand National Commission for UNESCO, Wellington, 1973, 367–371.
Fontaine, Y. A. (1974). In "Biochemical and Biophysical Perspectives in Marine Biology." (D. C. Malins and J. R. Sargent, Eds) pp. 139–212. Academic Press, London and New York.
Foote, J. S. (1916). *Smithson. Contr. Knowl.* **35**, 1–242.
Forrest, J. N., MacKay, W. C., Gallagher, B. and Epstein, F. H. (1973). *Am. J. Physiol.* **224**, 714–717.
Forte, J. G. and Nauss, A. H. (1966). *Am. J. Physiol.* **211**, 239–242.
Frazer, D. R. and Kodicek, E. (1970). *Nature* **228**, 764–766.
Frazier, L. W. (1976). *J. membrane Biol.* **30**, 187–196.
Fujita, T., Orimo, H., Yoshikawa, M., Morii, H. and Nakao, K. (1965). *Endocrinology* **76**, 1202–1204.
Garabedian, M., Holick, M. F., De Luca, H. F. and Boyle, I. T. (1972). *Proc. Nat. Acad. Sci.* Washington **69**, 1673–1676.
Garrod, D. and Newell, B. S. (1958). *Nature* **181**, 1411–1412.
Glimcher, M. J. (1976). In "Handbook of Physiology." (R. O. Greep, E. B. Astwood and G. D. Aurbach, Eds) Vol. 7, pp. 25–116. Williams and Wilkins, Baltimore.
Gray, T. K., Cooper, C. W. and Munson, P. L. (1974). In "M.P.P. International Review of Science." (S. M. McCann, Ed.). Physiology Series 1, **5**, pp. 239–275.
Guardabassi, A. (1960). *Z. Zellforsch. mikroskop. Anat.* **51**, 278–282.
Habener, J. F., Potts, J. T. and Rich A. (1976). *J. biol. Chem.* **251**, 3893–3899.
Halstead, L. B. (1974). "Vertebrate Hard Tissues." Wykenham, London.
Hamilton, D. W. (1964). *J. Morph.* **115**, 255–272.
Hamilton, J. W., MacGregor, R. R., Chu, L. L. H. and Cohn, D. V. (1971). *Endocrinology* **89**, 1440–1447.
Harell, A., Binderman, I. and Rodan, G. (1973). *Endocrinology* **92**, 550–555.
Hay, A. W. M. and Watson, G. (1977). In "Vitamin D. Biochemical, Chemical and Clinical Aspects Related to Calcium Metabolism." (A. W. Norman, K. Schaefer, J. W. Coburn, H. F. De Luca, D. Frazer, H. G. Grigaleit and D. V. Herrath, Eds) pp. 483–489. Walter de Gruyter, Berlin and New York.
Hayslett, J. P., Epstein, M., Spector, D., Myers, J. D., Murdaugh, H. V. and Epstein, F. H. (1971). *Bull. Mt. Desert. Is. Biol. Lab.* **11**, 33–35.

Heersche, J. N. M., Marcus, R. and Aurbach, G. D. (1974). *Endocrinology* **94**, 241–247.
Hellman, D. E., Au, W. Y. W. and Bartter, F. C. (1965). *Am. J. Physiol.* **209**, 643–650.
Henry, H. and Norman, A. W. (1975). *Comp. biochem. Physiol.* **50B**, 431–434,
Henry, H. L., Norman, A. W., Taylor, A. N., Hartenblower, D. L. and Coburn, J. W. (1976). *J. Nutr.* **106**, 724–734.
Hertelendy, F. and Taylor, I. G. (1961). *Poult. Sci.* **40**, 108–114.
Hess, A. F., Bills, C. E., Weinstock, M. and Rivkin, H. (1928). *Proc. Soc. exp. biol. Med.* **25**, 349–350.
Hickman, C. and Trump, B. E. (1969). *In* "Fish Physiology." (W. S. Hoar and D. J. Randall, Eds) Vol. 1, pp. 91–239. Academic Press, New York and London.
Hirsch, P. F., Gauthier, G. F. and Munson, P. L. (1963). *Endocrinology* **73**, 244–252.
Hodges, R. D. (1970). *Ann. Biol. anim. Bioch. Biophys.* **10**, 255–275.
Hodges, R. D. and Gould, R. P. (1969). *Experientia* **25**, 1317–1319.
Holick, M. F., Schnoes, H. K., De Luca, H. F., Suda, T. and Cousins, R. J. (1971). *Biochemistry* **10**, 2799–2804.
Holmes, W. N. and Donaldson, E. M. (1969). *In* "Fish Physiology." (W. S. Hoar and D. T. Randall, Eds) Vol. 1, pp. 1–89. Academic Press, London and New York.
Horrobin, D. F. (1973). "Prolactin: Physiology and Clinical Significance." MTP Medical and Technical Publishing Co. Ltd., Lancaster.
Horrobin, D. F. (1974). "Prolactin." MTP Medical and Technical Publishing Co. Ltd., Lancaster.
Huefner, M. and Hesch, R. D. (1971). *Kin. Wochenschr.* **49**, 1149.
Hughes, B. O. (1972). *Br. Poult. Sci.* **13**, 485–493.
Hughes, B. O. and Wood-Gush, D. G. M. (1971). *Anim. Behav.* **19**, 490–499.
Hurwitz, S., Bar, A. and Cohen, I. (1973). *Am. J. Physiol.* **225**, 150–154.
Istin, M. (1974). *In* "Biochemical and Biophysical Perspectives in Marine Biology." (D. C. Malins and J. R. Sargent, Eds) pp. 1–68. Academic Press, London and New York.
Johnstone, C. G., Schmidt, R. S. and Johnstone, B. M. (1963). *Comp. biochem. Physiol.* **9**, 335–341.
Joshua, I. G. and Mueller, W. J. (1977). *Fedn. Proc.* **36**, 138.
Kennedy, J. W. and Talmage, R. V. (1971). *Endocrinology* **88**, 1203–1209.
Kenny, A. D. (1975). *In* "Calcium Regulating Hormones." Proceedings of 5th Parathyroid Conference, Oxford, 1974. (R. V. Talmage, M. Owen and J. A. Parsons, Eds) pp. 408–410. Excerpta Medica, Amsterdam.
Kenny, A. D. (1976). *Am. J. Physiol.* **230**, 1609–1615.
Kenny, A. D. and Dacke, C. G. (1974). *J. Endocr.* **62**, 15–23.
Kenny, A. D. and Dacke, C. G. (1975). *Wld. Rev. Nutr. Dietetics* **20**, 231–298.
Kenny, A. D. and Musacchia, X. J. (1976). *Comp. biochem. Physiol.* **54A**, 1–5.
Kenny, A. D., Ahearn, D. J. and Maher, J. F. (1976). *Biochem. Med.* **16**, 201–210.
Kenny, A. D., Baksi, S. N., Galli-Gallando, S. M. and Pang, K. P. T. (1977). *Fedn. Proc.* **36**, 4366.
Kenny, A. D., Boelkins, J. N., Dacke, C. G., Fleming, W. R. and Hanson, R. C. (1972). *In* "Endocrinology 1971." Proceedings III International Symposium (S. Taylor, Ed.) pp. 39–47. Heinemann, London.
Keutmann, H. T., Aurbach, G. D., Dawson, B. F., Niall, H. D., Deftos, L. J. and Potts, J. T. (1971). *Biochemistry* **10**, 2779–2787.

REFERENCES

Kobayashi, S., Yamada, J. Maekawa, K. and Ouchi, K. (1972). *Biomineralization* **6,** 84–90.
Kraintz, L. and Intscher (1969). *Can. J. physiol. Pharmacol.* **47,** 313–315.
Krishnamurthy, V. G. and Bern, H. A. (1969). *Gen. comp. Endocr.* **13,** 313–335.
Kyes, P. and Potter, T. S. (1934). *Anat. Rec.* **60,** 377–379.
Leonard, F., Boke, J. W., Ruderman, R. J. and Hegyebo, A. F. (1972). *Calc. Tissue Res.* **10,** 269–279.
Levinsky, N. G. and Davidson, D. G. (1957). *Am. J. Physiol.* **191,** 530–536.
Lewis, P. E. and Taylor, T. G. (1972). *J. Endocr.* **53,** xlv–xlvi.
Livingstone, D. M. and Wacker, W. E. C. (1976). In "Handbook of Physiology." (R. O. Greep, E. B. Astwood and G. D. Aurbach, Eds) Vol. 7, pp. 215–224. Williams and Wilkins, Baltimore.
Lopez, E. (1970a). *Z. Zellforsch.* **109,** 552–565.
Lopez, E. (1970b). *Z. Zellforsch.* **109,** 566–572.
Lopez, E., Deville, J. and Bagot, E. (1968). *C. r. Acad. Sci.* Paris **267,** 1531–1534.
Lopez, E., Chartier-Baraduc, M. and Deville, J. (1971). *C. r. hebd. Séanc. Acad. Sci.* Paris **272,** 2600–2603.
Lopez, E., Peignoux-Deville, J., Lallier, F., Martelly, E. and Milet, C. (1976). *Calcif. Tiss. Res.* **20,** 173–186.
Lorcher, K. and Hodges, R. D. (1969). *Comp. biochem. Physiol.* **28,** 119–128.
Louw, G. N., Sutton, W. S. and Kenny, A. D. (1967). *Nature* **215,** 888–889.
Ma, S. W. Y. and Copp, D. H. (1978). In "Endocrinology of Calcium Metabolism." Proceedings of VI Parathyroid Conference, Vancouver, 1977. (D. H. Copp and R. V. Talmage, Eds). Excerpta Medica, Amsterdam and New York.
Ma, S. W. Y., Shami, Y., Messer, H. H. and Copp, D. H. (1974). *Biochem. biophys. Acta* **345,** 243–251.
McFarland, W. N. and Munz, F. W. (1965). *Comp. biochem. Physiol.* **14,** 383–398.
MacGregor, R. R., Chü, L. L. H., Hamilton, J. W. and Cohn, D. V. (1973). *Endocrinology* **92,** 1312–1317.
McWhinnie, D. J. (1975). *Comp. biochem. Physiol.* **50A,** 169–175.
McWhinnie, D. J. and Cortelyou, J. R. (1967). *Am. Zoologist* **7,** 857–868.
McWhinnie, D. J. and Lehrer, L. (1972). *Comp. biochem. Physiol.* **43A,** 911–925.
MacIntyre, I., Colston, K. W., Evans, I. M. A., Lopez, E., MacAuley, S. J., Peignoux-Deville, J., Spanos, E. and Szelke, M. (1976). *Clin. Endocrinol.* Suppl. **5,** 85s–95s.
Maetz, J. (1968). "Perspectives in Endocrinology. Hormones in the Lives of Lower Vertebrates." (E. J. W. Barrington and C. B. Jorgensen, Eds) pp. 47–162. Academic Press, London and New York.
Maetz, J. (1974). In "Biochemical and Biophysical Perspectives in Marine Biology." (D. C. Malins and J. R. Sargent, Eds) pp. 1–167. Academic Press, London and New York.
Martin, T. J., Vakaksis, N., Eisman, J. A., Livesey, S. J. and Tregear, G. W. (1974). *J. Endocr.* **63,** 369–375.
Martindale, L. (1969). *J. Physiol.* **203,** 82–83.
Marx, S. J., Woodward, C. J. and Aurbach, G. D. (1972). *Science* **178,** 999–1001.
Mashiko, K. and Jozuka, K. (1964). *Annotnes. zool. jap.* **37,** 41–50.
Mashiko, K., Jozuka, K. and Morita, O. (1964). *Ann. Rep. Noto. Mar. Lab. Univ. Kanazawa.* **4,** 53–58.
Melick, C. E., Aurbach, G. D. and Potts, J. T. (1965). *Endocrinology* **77,** 198–202.
Milhaud, G., Du, L. and Perault-Staub, A. M. (1971). *Rev. Eur. Etud. clin. biol.* **16,** 451–454.

Mongin, P. and Lacassagne, L. (1964). *C. r. hebd. Séanc. Acad. Sci.* Paris **258**, 3093-3094.
Moore, W. S. (1972). In "The Encyclopaedia of Geochemistry and Environmental Sciences." (R. W. Fairbridge, Ed.) pp. 208-213. Van Nostrand Reinhold Co., New York.
Morii, H and De Luca, H. F. (1967). *Am. J. Physiol.* **213**, 358-362.
Morris, T. R. (1973). *Poult. Sci.* **52**, 423-445.
Moss, M. L. (1961). *Acta Anat.* **46**, 343-362.
Moss, M. L. (1962). *Acta Anat.* **48**, 46-60.
Moss, M. L. (1963). *Ann. N.Y. Acad. Sci.* **109**, 337-350.
Moss, M. L. (1965). *Acta. Anat.* **60**, 262-276.
Mueller, W. J. and Leach, R. M. (1974). *Ann. Rev. Pharmacol.* **14**, 289-303.
Mueller, W. J., Hall, K. L., Maurer, C. A. and Joshua, I. G. (1973). *Endocrinology* **92**, 853-856.
Mugiya, Y. and Watabe, N. (1977). *Comp. biochem. Physiol.* **57A**, 197-202.
Munson, P. L. (1976). In "Handbook of Physiology." (R. O. Greep, E. B. Astwood and G. D. Aurbach, Eds) Vol. 7, pp. 443-464. Williams and Wilkins, Baltimore.
Munz, F. W. and McFarland, W. N. (1964). *Comp. biochem. Physiol.* **13**, 381-400.
Neuman, W. F. (1969). *Fedn Proc.* **28**, 1864-1850.
Neuman, W. F. and Mulryan, B. J. (1968). *Calc. Tiss. Res.* **2**, 237-241.
Neuman, W. F. and Ramp, W. K. (1971). In "Cellular Mechanisms for Calcium Transfer and Homeostasis." (G. Nichols and R. H. Wasserman, Eds) pp. 197-209. Academic Press, New York and London.
Neuman, W. F., Mulryan, B. J., Neuman, M. W. and Lane, K. (1973). *Am. J. Physiol.* **224(3)**, 600-606.
Niall, H. D., Keutmann, H. T., Copp, D. H. and Potts, J. T. (1969). *Proc. Nat. Acad. Sci.* Washington **64**, 771-778.
Nicoll, C. S. and Bern, H. A. (1972). In "Lactogenic Hormones 1972." pp. 299-317. Churchill Livingstone, London.
Nicoll, C. S., Meites, J. and Blackwell, C. (1962). *Endocrinology* **70**, 272-277.
Nieto, A., Fando, J. J. L. and Candela, J. L. R. (1975). *Gen. comp. Endocr.* **25**, 259-263.
Norman, A. W. and Henry, H. (1974). *Clin. Orthop. Rel. Res.* **98**, 258-286.
Ogawa, M. (1968). *Can. J. Zool.* **46**, 669-676.
Oguri, M. (1973). *Bull. Jap. Soc. scient. Fish.* **39**, 851-858.
Oguro, C. (1970). *Gen. comp. Endocr.* **15**, 313-319.
Oguro, C. and Uchiyama, M. (1975). *Gen. comp. Endocr.* **27**, 531-537.
Ohehy, D. A., Schmitt, R. A. and Bethard, W. F. (1966). *J. Nucl. Med.* **7**, 917-927.
Olivereau, M. and Chartier-Baraduc, M. M. (1966). *Gen. comp. Endocr.* **7**, 27-36.
Olson, E. B., De Luca, H. F. and Potts, J. T. (1972a). In "Calcium, Parathyroid Hormone and the Calcitonins." (R. V. Talmage and P. L. Munson, Eds) p. 240. Excerpta Medica, Amsterdam and New York.
Olson, E. B., De Luca, H. F. and Potts, J. T. (1972b). *Endocrinology* **90**, 151-157.
Omdahl, J. L. and De Luca, H. F. (1973). *Physiol. Rev.* **53**, 327-372.
Orimo, H., Ohata, M., Yoshikawa, M., Abe, J., Watanabe, S., Katani, M. and Higashi, T. (1971). *Igoku No Ayumi* **79**, 480.
Orimo, H., Ohata, M., Fujita, T., Yoshikawa, M., Higashi, T., Abe, J., Watanabe, S. and Otani, K. (1972). In "Endocrinology 1971." Proceedings of III International Symposium, London, 1971. (S. Taylor, Ed.) pp. 48-54. Heinemann' London.
Otani, M., Noda, T., Yamauchi, H., Watanabe, S., Matsuda, T., Orimo, H. and

REFERENCES

Narita, K. (1975). *In* "Calcium Regulating Hormones." Proceedings of 5th Parathyroid Conference, Oxford, 1974. (R. V. Talmage, M. Owen and J. A. Parsons, Eds) pp. 111–115. Excerpta Medica, Amsterdam and New York.
Pang, P. K. T. (1971a). *J. exp. Zool.* **178**, 15–22.
Pang, P. K. T. (1971b). *J. exp. Zool.* **178**, 89–100.
Pang, P. K. T. (1973). *Am. Zool.* **13**, 775–792.
Pang, P. K. T. and Pickford, G. E. (1967). *Comp. biochem. Physiol.* **21**, 573–578.
Pang, P. K. T., Clark, N. B. and Thomson, K. S. (1971a). *Gen. comp. Endocr.* **17**, 582–585.
Pang, P. K. T., Griffith, R. W. and Pickford, G. E. (1971b). *Proc. Soc. exp. biol. Med.* **130**, 85–87.
Pang, P. K. T., Pang, R. K. and Sawyer, W. H. (1973). *Endocrinology* **93**, 705–710.
Pang, P. K. T., Schreibman, M. P. and Griffith, R. W. (1973). *Gen. comp. Endocr.* **21**, 536–542.
Pang, P. K. T., Pang, R. K. and Sawyer, W. H. (1974). *Endocrinology* **94**, 548–555.
Parsons, J. A. (1976). *In* "The Biochemistry and Physiology of Bone." (G. H. Bourne, Ed.) Vol. IV, pp. 159–225. Academic Press, New York and London.
Parsons, J. A. and Robinson, C. J. (1971). *Nature* **230**, 581–582.
Parsons, J. A., Neer, R. M. and Potts, J. T. (1971). *Endocrinology* **89**, 735–740.
Parsons, J. A., Reit, B. and Robinson, C. J. (1973). *Endocrinology* **92**, 454–462.
Parsons, J. A., Gray, D., Rafferty, B. and Zanelli, J. (1978). *In* "Endocrinology of Calcium Metabolism." Proceedings of 6th Parathyroid Conference, Vancouver, 1977. (D. H. Copp and R. V. Talmage, Eds) pp. 111–114. Excerpta Medica, Amsterdam and New York.
Pautard, F. (1961). *New Scientist* **12**, 364–366.
Pearse, A. G. E. (1968). *Proc. Roy. Soc.* **170**, Series B, 71–80.
Pearse, A. G. E. (1976). *In* "Handbook of Physiology." (R. O. Greep, E. B. Astwood and G. D. Aurbach, Eds) Vol. 7, pp. 411–422. Williams and Wilkins, Baltimore.
Pearse, A. G. E. and Taylor, T. T. (1976). *Clin. Endocrinol.* Suppl. **5**, 229S–244S.
Peignoux-Deville, J., Milet, C. and Martelly, E. (1978). *Ann. Biol. anim. biochem. Biophys.* **18**, 119–126.
Phillipo, M., Bates, R. F. L. and Lawrence, C. B. (1971). *J. Endocr.* **49**, vi–vii.
Pike, J. W. and Alvarado, R. H. (1975). *Comp. biochem. Physiol.* **51B**, 119–125.
Pilkington, J. B. and Simkiss, K. (1966). *J. exp. Biol.* **45**, 329–341.
Pitts, R. F. (1968) (2nd edn). "Physiology of the Kidney and Body Fluids." Year Book. Medical Publishers Inc., Chicago.
Polin, D. and Sturkie, P. D. (1958). *Endocrinology* **63**, 177–182.
Polin, D., Sturkie, P. D. and Hunsacker, W. (1957). *Endocrinology* **60**, 1–5.
Ponchon, G., Kennan, A. L. and De Luca, H. F. (1969). *J. clin. Invest.* **48**, 2023–2037.
Poston, H. A. (1969). *Fisheries Bull.* **22**, 48–50.
Potts, J. T. (1976). *In* "Peptide Hormones." (J. A. Parsons, Ed.) pp. 119–143. Macmillan, London.
Potts, J. T. and Aurbach, G. D. (1976). *In* "Handbook of Physiology." (R. O. Greep, E. B. Astwood and G. D. Aurbach, Eds) Vol. 7, pp. 423–430. Williams and Wilkins, Baltimore.
Potts, J. T., Niall, H. D., Keutmann, H. T. and Lequin, R. M. (1971). *In* "Calcium, Parathyroid Hormone and Calcitonin." (R. V. Talmage and P. L. Munson, Eds) p. 121. Excerpta Medica, Amsterdam.

Potts, J. T., Keutmann, H. T., Niall, H. D., Habener, J. F. and Tregear, G. W. (1972). *Gen. comp. Endocr.* Suppl. **3**, 405–410.
Prien, E. L., Pyle, E. B. and Krane, S. M. (1976). In "Handbook of Physiology." (R. O. Greep, E. B. Astwood and G. D. Aurbach, Eds) Vol. 7, pp. 383–410. Williams and Wilkins, Baltimore.
Radde, I., Hoffken, B., Parkinson, D. K., Sheepers, J. and Luckham, A. (1971). *Clin. Chem.* **17**, 1002–1006.
Raisz, L. G. (1976). In "Handbook of Physiology." (R. O. Greep, E. B. Astwood and G. D. Aurbach, Eds) Vol. 7, pp. 117–136. Williams and Wilkins, Baltimore.
Rankin, J. C., Chan, D. K. O. and Chester-Jones, I. (1967). *Gen. comp. Endocr.* **9**, 484–485.
Rasmussen, H. and Westall, R. G. (1956). *Nature* **178**, 1173–1174.
Rasmussen, H., Wong, M., Bikle, D., Goodman, D. B. P. (1972). *J. clin. Invest.* **51**, 2502–2504.
Rasmussen, H., Goodman, D. B. P., Friedmann, N., Allen, J. E. and Kurakawa, K. (1976). In "Handbook of Physiology." (R. O. Greep, E. B. Astwood and G. D. Aurbach, Eds) Vol. 7, pp. 225–264. Williams and Wilkins, Baltimore.
Rasquin, P. and Rosenbloom, L. (1954). *Bull. Am. Mus. nat. Hist.* **104**, 363–425.
Rawson, D. S. (1939). In "Problems of Lake Biology." (F. R. Moulton, Ed.). Lancaster, Pennsylvania, A.A.A.S.
Reid, D. F., Ego, W. T. and Townsley, S. J. (1959). *Anat. Rec.* **134**, 628.
Reynolds, J. J., Dingle, J. T., Gudmundsson, T. V. and MacIntyre, I. (1968). In "Calcitonin." Proceedings of Symposium Thiprocalcitonin and the C Cells (1967). (See Capp *et al.*, 1968) (S. Taylor and R. Frazer, Eds) pp. 223–229. Heinemann, London.
Robertson, D. R. (1965). *Z. Zellforsch.* **67**, 584–599.
Robertson, D. R. (1967). *Trans. Am. Micr. Soc.* **86**, 195–203.
Robertson, D. R. (1968a). *Z. Zellforsch.* **90**, 273–288.
Robertson, D. R. (1968b). *Z. Zellforsch. mikrosk. Anat.* **85**, 441–542.
Robertson, D. R. (1969a). *Gen. comp. Endocr.* **12**, 479–490.
Robertson, D. R. (1969b). *Endocrinology* **84**, 1174–1178.
Robertson, D. R. (1969c). *J. exp. Zool.* **172**, 425–442.
Robertson, D. R. (1970). *Endocrinology* **87**, 1041–1050.
Robertson, D. R. (1971). *Gen. comp. Endocr.* **16**, 329–341.
Robertson, D. R. (1972a). In "Calcium, Parathyroid Hormone and the Calcitonins." Proceedings of 4th Parathyroid Conference, Chapel Hill, 1971. (R. V. Talmage and P. L. Munson, Eds) pp. 21–28. Excerpta Medica, Amsterdam and New York.
Robertson, D. R. (1972b). *Gen. comp. Endocr.* Suppl. **3**, 421–429.
Robertson, D. R. (1974). *Endocrinology* **94**, 940–946.
Robertson, D. R. (1975a). *Comp. biochem. Physiol.* **51A**, 705–710.
Robertson, D. R. (1975b). *Endocrinology* **96**, 934–940.
Robertson, D. R. (1976). *Comp. biochem. Physiol.* **54A**, 225–231.
Robertson, D. R. (1977). *Gen. Comp. Endocr.* **33**, 336–343.
Romer, A. S. (1960). "Vertebrate Paleontology." University Press, Chicago.
Romer, A. S. (1963) (3rd edn). "The Vertebrate Body." W. B. Saunders Co., Philadelphia and London.
Romer, A. S. (1964). In "Bone Biodynamics." (H. M. Frost, Ed.) pp. 13–40. Little Brown and Co., Boston.
Roos, B. A., Okano, K. and Deftos, L. J. (1974). *Biochem. biophys. Res. Comm.* **60**, 1134–1140.

Roth, S. I. and Schiller, A. L. (1976). *In* "Handbook of Physiology." (R. O. Greep, E. B. Astwood and G. D. Aurbach, Eds) Vol. 7, pp. 281–312. Williams and Wilkins, Baltimore.
Russell, R. G. G., Casey, P. A. and Fleisch, H. (1968). *Calcif. Tiss. Res.* Suppl. **2**, 54–54A.
Ruth, E. S. (1918). *Philipp. J. Sci.* **13**, 311–318.
Schlumberger, H. G. and Burk, D. H. (1953). *Arch. Path.* **56**, 103–124.
Schmidt, R. S. (1963). *Comp. biochem. Physiol.* **10**, 83–87.
Schraer, H., Hohman, W., Ehrespeck, G. and Schraer, R. (1965). *J. cell. Biol.* **27**, 96A.
Schraer, R. and Schraer, H. (1965). *Proc. Soc. exp. Biol. Med.* **119**, 937–942.
Sebell, W. H. and Harris, R. S. (1954). *In* "The Vitamins." pp. 131–266. Academic Press, London and New York.
Sedrani, S. and Taylor, T. G. (1977). *J. Endocr.* **72**, 405–406.
Senior, B. E. (1973). Oestrogens in the Domestic Fowl, Ph.D. thesis, University of Reading, U.K.
Shapiro, H. A. and Zwarenstein, H. (1933). *J. exp. Biol.* **10**, 186–195.
Sherwood, L. M., Mayer, G. P., Romberg, C. F., Kronfield, D. S., Potts, J. T. and Aurbach, G. D. (1966). *J. clin. Invest.* **45**, 1072–1073.
Sherwood, L. M., Abe, M., Rodman, J. S., Lundberg, W. B. and Targovnik, J. H. (1972). *In* "Calcium, Parathyroid Hormone and the Calcitonins." Proceedings of 4th Parathyroid Conference, Chapel Hill, 1971. (R. V. Talmage and P. L. Munson, Eds) pp. 183–196. Excerpta Medica, Amsterdam and New York.
Simkiss, K. (1961). *Biol. Rev.* **36**, 321–376.
Simkiss, K. (1967). "Calcium in Reproductive Physiology." Chapman and Hall, London.
Simkiss, K. (1968). *Am. J. Physiol.* **214**, 627–634.
Simkiss, K. (1974). *In* "Ageing of Fish." (T. B. Bagenal, Ed.) pp. 1–12. Unwin Brothers, Old Woking.
Simkiss, K. (1975). *Symp. zool. Soc.* London **35**, 307–337.
Simkiss, K. and Dacke, C. G. (1971). *In* "Physiology and Biochemistry of the Domestic Fowl." (D. J. Bell and B. M. Freeman, Eds) pp. 481–488. Academic Press, London and New York.
Simmons, D. J. (1971). *Clin. Orthop. Rel. Res.* **76**, 244–280.
Simmons, D. J. (1976). *In* "The Biochemistry and Physiology of Bone." (G. H. Bourne, Ed.) Vol. IV, pp. 445–516. Academic Press, New York and London.
Simmons, D. J., Simmons, N. B. and Marshall, J. H. (1970). *Calc. Tiss. Res.* **5**, 206–221.
Simmons, D. J., Hakim, R. and Cummins, H. (1971). *Experientia* **27**, 1210–1211.
Smith, H. M. (1964). "Evolution of Chordate Structure." Holt, Rinehart and Winston, New York.
Spanos, E., Colston, K. W., Evans, I. M. S., Galante, L. S., Macauley, S. J. and MacIntyre, I. (1976). *Molec. cell. Endocr.* **5**, 163–167.
Speers, G. M., Perey, D. Y. E. and Brown, D. M. (1970). *Endocrinology* **87**, 1292–1297.
Stanley, J. C. and Fleming, W. R. (1964). *Science* **144**, 63–64.
Studitsky, A. N. (1945). *C. r. Acad. Sci.* **47**, 444–447.
Sulze, W. (1942). *Arch. Ges. Physiol.* **246**, 250–257.
Talmage, R. V. (1972). *In* "Calcium, Parathyroid Hormone and the Calcitonins." Proceedings of 4th Parathyroid Conference, Chapel Hill, 1971. (R. V. Talmage and P. L. Munson, Eds) pp. 422–429. Excerpta Medica, Amsterdam and New York

Talmage, R. V. and Meyer, R. A. (1976). In "Handbook of Physiology." (R. O. Greep, E. B. Astwood and G. D. Aurbach, Eds) Vol. 7, pp. 313–342. Williams and Wilkins, Baltimore.
Talmage, R. V., Anderson, J. J. B. and Cooper, C. W. (1972). *Endocrinology* **90**, 1185–1191.
Talmage, R. V., Roycroft, J. H. and Anderson, J. J. B. (1975). *Calc. Tiss. Res.* **17**, 91–102.
Taylor, T. G. (1970). *Ann. Biol. anim. Biochem. Biophys.* **10**, 83–91.
Taylor, T. G. (1971). In "Physiology and Biochemistry of the Domestic Fowl." (D. J. Bell and B. M. Freeman, Eds) Vol. 1, pp. 473–480. Academic Press, London and New York.
Taylor, T. G. and Hertelendy, F. (1961). *Poult. Sci.* **40**, 115–123.
Taylor, T. G. and Moore, J. H. (1954). *Br. J. Nutr.* **8**, 112–124.
Thorson, T. B. (1961). *Biol. Bull.* **120**, 238–254.
Toribara, T. Y., Terepka, A. R. and Dewey, P. A. (1957). *J. clin. Invest.* **36**, 738.
Torrey, T. W. (1971) (3rd edn). "Morphogenesis of the Vertebrates." John Wiley, New York, London, Sydney and Toronto.
Tregear, G. W., Rietschoten, J. V., Greene, E., Keutmann, H. T., Niall, H. D., Reit, B., Parsons, J. A. and Potts, J. T. (1973). *Endocrinology* **93**, 1349–1353.
Triffitt, J. T., Terepka, A. R. and Neuman, W. F. (1968). *Calc. Tiss. Res.* **21**, 165–176.
Umansky, E. E. and Kudokatzev, V. P. (1951). *Dokl. Akad. Nauk. SSSR* **77**, 533.
Urist, M. R. (1961). *Endocrinology* **69**, 778/801.
Urist, M. R. (1962). *Perspec. Biol. Med.* **6**, 75–115.
Urist, M. R. (1963). *Proc. N.Y. Acad. Sci.* **109**, 294–311.
Urist, M. R. (1964). In "Bone Biodynamics." (H. M. Frost, Ed.) pp. 151–179. Little Brown and Co., Boston.
Urist, M. R. (1967). *Am. Zool.* **7**, 883–895.
Urist, M. R. (1973). *Comp. biochem. Physiol.* **44A**, 131–135.
Urist, M. R. (1976a) (2nd edn). In "The Biochemistry and Physiology of Bone." (G. H. Bourne, Ed.) Vol. IV, pp. 1–59. Academic Press, London and New York.
Urist, M. R. (1976b). In "Handbook of Physiology." (R. O. Greep, E. B. Astwood and G. D. Aurbach, Eds) Vol. 7, pp. 183–213. Williams and Wilkins, Baltimore.
Urist, M. R., Deutsch, N. M., Pomerantz, G. and McLean, F. C. (1960). *Am. J. Physiol.* **199**, 851–855.
Urist, M. R., Uyeno, S., King, E., Okada, M. and Applegate, S. (1972). *Comp. biochem. Physiol.* **42A**, 393–408.
Van Oosten, J. (1957). "The Physiology of Fishes." (M. E. Brown, Ed.) Vol. 1, pp. 207–244. Academic Press, London and New York.
Vaughan, J. M. (1970). "The Physiology of Bone." Clarendon Press, Oxford.
Wachman, A. and Bernstein, D. S. (1970). *Clin. Orthop. Rel. Res.* **69**, 252–263.
Waggener, R. A. (1929). *J. Morph.* **48**, 1–44.
Waggener, R. A. (1930). *J. exp. Zool.* **57**, 13–56.
Wallin, O. (1957). *Rep. Inst. Freshwat. Res.* Drottningholm **38**, 385–447.
Wasserman, R. H. and Corradino, R. A. (1971). *Ann. Rev. Biochem.* **40**, 501–532.
Wasserman, R. H. and Taylor, A. N. (1976). In "Handbook of Physiology." (R. O. Greep, E. B. Astwood and G. D. Aurbach, Eds) Vol. 7, pp. 137–156. Williams and Wilkins, Baltimore.
Watlington, C. O., Burke, P. K. and Estep, H. L. (1968). *Proc. Soc. exp. Biol. N.Y.* **128**, 853–856.
Watts, E. G., Copp, D. H. and Deftos, L. J. (1975). *Endocrinology* **96**, 214–218.

Weber, J. C., Pons, V. and Kodicek, E. (1971). *Biochem. J.* **125**, 147–153.
Wener, J. A., Gorton, S. J. and Raisz, L. G. (1972). *Endocrinology* **90**, 752–759.
Whiteside, B. (1922). *Am. J. Anat.* **30**, 231–266.
Whitfield, M. (1977). *In* "Environmental Physiology of Animals." (J. Bligh, J. L. Cloudsley-Thompson and A. G. Macdonald, Eds) pp. 30–45. Blackwell, Oxford and London.
Williams, R. J. P. (1970). *Q. Rev. Chem. Soc.* London **24**, 331–365.
Wills, M. R. (1970). *Lancet* **2**, 802–804.
Wong, R. G., Myrtle, J. F., Tsai, H. C. and Norman, A. W. (1972). *J. biol. Chem.* **247**, 5728–5735.
Woodard, A. E. and Mather, F. B. (1964). *Poult. Sci.* **43**, 1427–1432.
Woodhead, P. M. J. (1968). *J. mar. Biol. Ass.* U.K. **48**, 81–91.
Woodhead, P. M. J. (1969). *Gen. comp. Endocr.* **13**, 310–312.
Yalow, R. S. and Berson, S. A. (1966). *Trans. N.Y. Acad. Sci.* **28**, 1033–1044.
Yoshida, K. and Talmage, R. V. (1962). *Gen. comp. Endocr.* **2**, 551–557.
Ziegler, R., Delling, G. and Pfeiffer, E. F. (1970). *In* "Calcitonin 1969." Proceedings of 2nd International Symposium. (S. Taylor, Ed.) pp. 301–310. Heinemann, London.
Zwarenstein, H. and Shapiro, H. A. (1933). *J. exp. Biol.* **10**, 372–378.

Subject Index

A

Accretion
 definition, 14
 of hard tissues, 14, 16, 24, 26, 27, 32
Acellular bone, 27–29, 39
 as a phosphorus store, 29, 187
 in fish, 4, 27–29, 187
Acetazolamide, 176
Acid-base regulation, 7, 10
 evolution in air breathers, 7, 10, 124
 in amphibians, 141–146, 188–191
 in egg-laying hen, 176–179
 interaction with Ca metabolism, 188–194
Acid excretion
 by toad bladder *in vitro*, 149, 191
 by avian oviduct, 177, 178
Acid phosphatase, 168
Acidosis
 bone carbonate mobilization in, 190–193
 in egg-laying hens, 177, 178
 metabolic, 177, 178, 190–194
 respiratory, 142, 143, 188, 189, 194
ACTH, *see* Corticotrophin
Actinomycin, 78
Adenosine triphosphate, 3, 16, 81
Adenyl cyclase, 48, 77, 158, 162
Adrenal interaction with calcitonin
 effect on kidney, 83
Adrenalectomy, 83
Ahemeral light cycles, 175
Albumen, 66
Alkaline earths
 choice of, in mineralized tissues, 12–14
 physico-chemistry of, 13, 14
Alkaline phosphatase, 16, 18, 34, 76, 77, 161, 168
 blood levels in hens, 168
 gut activity of, 80
 in shark blood, 18
 in shark cartilage, 34
 inhibition of calcification, 16
 response to calcitonin, 76, 161
 response to 1,25-dihydroxycholecalciferol, 80
 response to parathyroid hormone, 75
Alkalosis, 145, 194
Ammonia production by avian oviduct, 178
Amorphous Ca phosphate, 15
Androgen
 medullary bone formation and, 33, 78, 165, 181
 osteichthyan response to, 121, 122
Angiotensin, 68, 113
An-ionic skeleton, *see* Silica-based skeletons
Anti-oestrogens and vitamin D_3 metabolism, 174
Apatite, *see also* hydroxyapatite, 12, 15, 21, 22, 34, 35
Apposition, 14–16
Ascorbic acid, *see* Vitamin C
Aspidin, 39
Atmospheric CO_2 influence on acid-base regulation, 141–145, 194
ATP, *see* Adenosine triphosphate, *see also* ATPase
Ca^{2+}-ATPase, 16, 86, 88, 113
Na^+-ATPase, 86, 120

B

Balance studies in birds, 165
Bicarbonate
 as a blood buffer, 189, 190
 freshwater concentration of, 9, 10
 renal excretion of, 82, 83, 190
 sea water concentration, 6, 7
Bioassay, 46–48, 51, 57, 59–61, 67, 68, 70, 106, 108, 113
Blood bicarbonate
 in fish, 189

SUBJECT INDEX

Blood bicarbonate—*cont'd*
  in tetrapods, 189
Blood PCO_2
  in fish, 189
  in tetrapods, 189
Blood pH
  in fish, 189
  in tetrapods, 189
Bone, *see also* Medullary bone, Skeleton
  as a hormonal target organ, 74–79
  as a phosphorus store, 3, 4, 183, 185–187
  carbonate, 12, 21, 22, 35, 146, 149, 188–193
  cells, *see also* Osteoblast, Osteoclast, Osteocyte, 23–26
  CO_2, 22, 189
  density
    changes in reproducing reptiles, 155
    response to calcitonin, 76, 152, 161, 170, 171
  embryonic, *see* Embryonic bone
  evolution of, 3, 4, 183, 185, 187
  exchangeable fraction of, 14
  extracellular fluid of, 22, 23, 37
  matrix, 3, 12, 14, 15–18, 21, 24, 26, 30, 33, 39
    response to parathyroidectomy, 128
  membrane, 22, 23
  methods of formation, 26, 27
  remodelling of, 25, 27, 28
  resorption of, 16, 24, 25, 27, 30, 125, 159, 161, 164, 167, 168
Breeding cycles, 100 121
Buffer base reserve in the inner ear, 38, 145, 146, 188, 189

C

^{14}C-bicarbonate as bone label in acid-base studies, 190
C cells, 52, 53, 159, 162
Calcification, *see* Mineralization
Calcified cartilage, 17, 34
  in primitive vertebrates, 34
Calciolysis, 14
Calcitonin, 5, 51–62
  age dependency of action, 77
  amino-acid sequence of, 54–56
  amphibian response to, 132
  assay of, 57, 106, 108
  avian responses to, 160, 161, 170–172
  bone response to, 76, 77, 184–186
    in amphibians, 130–132
    in birds, 160, 161, 170, 171
    in osteichthyans, 101, 103, 104, 109, 110
    in reptiles, 152
  chemistry of, 53–57
  chondrichthyan response to, 98
  chronic responses to, 104, 151, 152, 161, 170, 171
  circulating level of, 58–62, 106, 108, 109, 169–171, 186
  cyclic AMP mediation of action in bone, 76, 77
  degradation of, 59
  discovery of, 51, 52
  eggshell thickness response to, 171, 172
  evolution of, 185
  extraction and concentration from plasma, 60, 61
  fish scale response to, 103, 104
  gill response to, 87, 88, 103
  gut response to, 80, 137, 175
  heterologous species of, 57
  natriuretic effect of, 83
  osmoregulatory response of fish to, 102, 103, 106, 107, 110
  osteoblast response to, 76
  osteoclast response to, 76, 161
  osteocyte response to, 76, 161
  osteichthyan response to, 87, 100–110
  precursor of, *see* Pro-calcitonin
  renal response to, 83, 130, 131
  reptilian response to, 151, 152
  response of hens to, 170–172
  role in phosphate regulation, 184–186
  salmon, 54–57, 59, 102, 109
  seasonal response by fish, 103, 104
  secretion of, 58, 59, 161–163, 186
    β-adrenergic influence, 161, 162
    Ca ion influence, 58
    3′, 5′-cyclic AMP role in, 58, 161
    gonadal hormone influence, 58
    gut hormone influence, 58, 186
  secretion rate, 161–163
  ultimobranchial content of, 54, 106, 108, 169, 170

SUBJECT INDEX 213

Calcitonin—*cont'd*
  pre-reproductive surge and phosphate regulation, 186
Ca adenosine 5′-triphosphatase *see* Ca^{2+} ATPase
Ca appetite, 179, 180
Ca
  binding protein
    activity in fish gills, 86, 87
    and vitamin D_3 action, 78, 80, 81, 88, 176
    in avian oviduct, 88, 89, 176
    in chondrichthyan plasma, 97
  body fluid concentration of, 10, 11
  cycle in nature, 7, 8, 10
  diffusible fraction in plasma, 10, 11
  freshwater concentration of, 9, 10
  in soft tissue, 1, 40
  ionized fraction in plasma, 10, 11
  level in reproducing female plasma, 147, 152–154
  level in bone fluid, 22
  physico-chemistry of, 12–14
  protein bound fraction in plasma, 10, 11
  ions role in biological processes, 1
  pumps, 4, 13, 24
    in bone cells, 74
    in gills, 86
    in gut, 79
    in oviduct, 88, 176
    in skin, 85
  sea water concentration of, 6, 7
  turnover
    in acellular bone, 28, 29, 103, 104, 117, 118, 187
    in fish scales, 30, 31, 103, 104, 117, 118, 187
  urine response to parathyroidectomy, 125
Ca-45
  and bone response to calcitonin, 77, 103, 104
  chloride studies in acid-base regulation, 176, 177
  flux studies in fish, 86–88, 102, 103, 117, 118
  gut uptake studies of, in birds, 175
  skin uptake by, in fish, 85
  in calcified cartilage, 34
  in fish scales, 31, 118, 119, 121
  in hens, 165, 166, 176
  parathyroid hormone response, 126, 157, 158
  bone response to calcitonin, 184, 185
  turnover in hypophysectomized fish, 117–119
  uptake by acellular bone, 28
  uptake by fish gills, 86
  uptake studies in fish, 85–88, 103, 104, 115–119
Calculi, *see* Renal stones
Calvariae, studies with, 22, 27, 77, 125
Canaliculi, 23, 27, 39
Canals of Williamson, 28
Carapace, of turtles, 31, 147
Carbohydrate metabolism
  in bone, 127, 128
  in hibernating animals, 194
Carbon dioxide
  dependency of parathyroid hormone action, 189
  partial pressure of, in blood, 142–145, 177, 189, 194
  in bone, 22, 189
  in eggshell formation, 176, 178
Carbonate-apatite, 22, 189
Carbonate
  as a blood buffer, 35, 38, 142, 146, 188–190
  as a buffer base reserve, 188–190
Carbonate of Ca
  in bone, 12, 21, 22, 35, 84, 146, 149, 188, 189, 190
  in eggshell, 40, 176
  in endolymphatic sacs, 35–38, 84, 149, 188, 190
  in the inner ear, 12, 35–38
Carbonic acid
  response to parathyroid hormone, 75
  in bone resorption, 75
Carbonic anhydrase
  and renal bicarbonate excretion, 83, 190
  in amphibian endolymphatic sacs, 37, 145
  in avian oviduct, 176–179
  inhibitors of, 176
  in inner ear, 37
  in kidney, 82, 83
  in bone response to parathyroid hormone, 189

γ-Carboxyglutamic acid (Gla), 17, 18
Carp pituitary extract, 104, 121
Carrier proteins
  for vitamin D_3 metabolites in blood, 66
  for Ca in blood, 10, 11
Cartilage, see also Calcified cartilage, 17, 26, 27
  calcification of, 26, 34
  chondrichthyan, 34, 96, 97
  cyclostome, 34, 94
  matrix of, 34
  nutrition of, 34
  in endochondral bone formation, 26, 27
Cartilaginous skeletons, 4
Castration
  effect on medullary bone formation, 33
  effect on plasma calcitonin level, 170
Cell membrane
  electrical properties, 1, 37, 40
Cellular membranes and Ca transfer, 165
Cellular organelles, role in calcification, 20
Cementum, 29
Cholecalciferol, see Vitamin D_3
Chondroblast, 34
Chondroclast, 34
Chondrocyte, 34
Chondroitin-sulphate, 17, 40
Circadian rhythm
  of acidosis, 177
  of Ca intake, 179
Citric acid
  response to calcitonin, 76
  response to parathyroid hormone, 75
Cleidoic eggs, 88
Closed system of Ca regulation, 12, 93
Collagen, 12, 16, 24
  Ca binding and, 15, 16
  chain types in, 15
  in dentine, 39
  in medullary bone, 33
Collagenolytic activity of bone, 168, 169
Compartmental considerations in Ca metabolism, 4
Corpuscles of Stannius, see also Stanniectomy, Hypocalcin, 67–69, 97
  assay of active substance, 67, 68, 113
  effect on plasma phosphate level, 111
  embryonic derivation, 67
  gill ATPase response to, 86, 113
  histology of, 68, 69, 105, 114
  histological response of, to ultimobranchialectomy, 105
  morphology, 67–69
  osmoregulatory role, 67, 110, 113, 114
  regulation of activity, 68, 69, 114
  in fish, 110–114
Cortical bone, 33, 168, 169
  chronic response to calcitonin, 170, 171
  response to Ca deficiency, 168
Corticotrophin, 44, 45, 69, 95
  osteichthyan response to, 120, 121
  plasma Ca response to, 120
Cortisol, 95, 97, 120
  interaction with calcitonin, 83
Crystal seeding, 15, 20 40
CT, see Calcitonin
Cyclic adenosine monophosphate see 3′,5′-cyclic AMP
3′,5′-Cyclic AMP
  in C cell, 161
  in kidney tubule, 82, 83
  in calcitonin secretion, 58
  in toad-bladder response to parathyroid hormone, 191

D

Dentine
  structure of, 38, 39
  in fish scales, 30, 39
  resorption of, 39
Deoxyribonucleic acid, see DNA
Dermal bone, 26, 27, 29, 31
  in chondrichthyans, 27
  in tetrapods, 31
Dermal denticles, 27, 96
Dibutyryl cyclic-AMP, 162
Dietary Ca uptake, 115, 123
1,25-Dihydroxycholecalciferol, 63–67
  assay of, in blood, 66, 67
  bone response to, 77, 78
  circulating levels of, 66, 67
  effect on bone protein synthesis, 78
  gut response, 80, 81, 135, 163, 164, 174
  plasma phosphate response of fish, 115, 187

SUBJECT INDEX 215

production in egg-laying hen, 172–174
renal response, 84
in Ca regulation, 195
secretion of, 65, 66
synthesis of, 63–66
  androgen influence, 174
  calcitonin influence, 65
  oestrogen influence, 66, 174
  parathyroid hormone influence, 65
  phosphate influence, 65
  prolactin influence, 65, 66
24,25-Dihydroxycholecalciferol, 64, 81, 172–174
25,26-Dihydroxycholecalciferol, 65
Diurnal periodicity, gut Ca uptake of, 138, 139
Diurnal variation, see also Circadian rhythm
  in plasma phosphate level, 186
DNA, 78
Drinking rate in fish, 113, 115, 116

E

Eggshell
  embryonic resorption of, 40, 166
  of megalecithal eggs, 40
  quality of, 164, 172, 181
  reptilian, 40, 152
  secretion of, 88, 89, 152, 166, 167, 169, 171, 172, 175, 179, 181
  structure of, 40
Eggshell thickness response to parathyroid hormone, 89, 166, 167
Electrical potential of endolymphatic membrane, 37
Embryo, Ca metabolism of, 88, 166, 196
Embryonic bone, see also Calvariae, Foetal bone, 26, 27
Enamel, 39
Enameloid, 39
Endochondral bone, 26, 27, 31
Endocrines, see specific hormones and endocrine glands
Endolymphatic sacs, see also Inner ear, 35–38
  and acid-base balance, 38, 141–146, 149, 188–190, 194
  changes in reproducing females, 154, 155

in embryo, 38
in reptiles, 154, 155
role in amphibian metamorphosis, 141
Endometrium, 179
Endosteal bone, 32, 78
Environmental Ca, see also Ca cycle in nature, 6–10
Environmental magnesium, see also Magnesium cycle in nature, 6–10
Environmental phosphorus, see also Phosphorus cycle in nature, 6–10
Enzymes, see specific enzymes
Escape phenomenon, 76, 171
Estrogen, see Oestrogen
Euryhaline fish, 90, 92
  magnesium homeostasis of, 188
  variations in plasma Ca level, 99, 100
  variations in plasma phosphate level, 100
Extracellular fluid volume, 100

F

Fluid
  extracellular, 1, 4, 6, 7, 9–11, 15, 18, 19, 22, 36, 37
  intracellular, 4, 13, 37
Foetal bone, see also Embryonic bone, calvariae, 27, 76
Fossil bone, 3, 25, 26, 27, 31, 38, 39
Fracture healing, 28, 132
Freshwater environment, composition of, 9, 10, 19

G

Gastrin effect on calcitonin secretion, 58
Gastrointestinal tract, see Gut
Gill, 86–88, 113, 120, 123
  ATPases in, 86, 113, 120
  Ca uptake by in fish, 86–88, 92, 93, 113, 116, 118, 120, 123
  role in acid-base balance, 145, 146
  role in hydromineral regulation, 86, 87, 120
Gla, see γ-carboxyglutamic acid
Globulin, 66
Glomerular filtration rate, 82, 125, 131
Glycosaminoglycan, 16, 17
Golgi bodies, 53, 69
Gonads, plasma calcitonin response to, 109

SUBJECT INDEX

Gonadal hormones, *see also* Oestrogens, Androgens, 70, 71
  and medullary bone formation, 33, 70
  in blood,
  pre-laying surge in hens, 170
  lack of effect on chondrichthyan plasma Ca levels, 97
  osteichthyan responses to, 121, 122
  reptilian responses to, 152
Growth hormone, *see* somatotrophin
Growth rate, effect of calcitonin, 151, 152
Gut, 79–81
  Ca absorption by, in egg-laying bird, 174, 175
  Ca uptake by, in amphibians, 134, 139
  Ca uptake by, in fish, 115, 116
  Ca uptake in Ca depleted birds, 175
  magnesium uptake by, 80
  response to calcitonin, 80, 135, 137, 175
  response to 1,25-dihydroxycholecalciferol, 80, 81, 115, 135, 164, 175, 187
  response to 24,25-dihydroxycholecalciferol, 81
  response to parathyroid hormone, 79, 80, 159

H
Haversian systems, 25
Hibernation, 53, 194
Hormones, *see also* specific hormones and endocrine glands, 41–72
  role in cyclostome, Ca regulation, 95
Howship's lacunae, 25
Hydrogen ion concentration (pH), *see also* acid-base regulation, 6, 7, 9–11, 16, 49, 75, 142–146, 177, 178, 189–194
  calcification and, 16, 177, 178
Hydrosphere, composition of, 8, 9
Hydroxyapatite, 12, 15, 21, 35
  chemical structure of, 21
  in calcified cartilage, 34
  in inner ear of cyclostomes, 38
25-hydroxycholecalciferol, *see also* Vitamin D_3, 63, 65, 66, 163
  negative feedback regulation of synthesis, 63
25-hydroxycholecalciferol-1-hydroxylase (renal), 64, 66, 174

Hydroxyproline
  urinary excretion of, 168
  parathyroid hormone effect on, 76, 159
  in acidosis, 191, 193
Hypercalcaemia in reproducing females, 11, 70, 71, 97, 98, 109, 121, 122, 147, 152, 153, 170, 171, 181
Hypercalcin, 70, 121, 195
Hypermagnesaemia in reproducing females, 153, 154
Hyperparathyroid disease, 190
Hyperphosphataemia in reproducing females, 153, 154
Hypocalcaemic response to calcitonin, 74, 76, 98, 101, 102, 132, 151, 161, 185, 186
Hypocalcin, 5, 67, 68, 97, 195
  assay methods for, 68, 113
  gill response to, 86, 87, 113
  osteichthyan responses to, 110–114
Hypophysectomy
  amphibian response to, 139, 140
  bone response to, 117–119
  effect on Ca turnover, 117–119
  osteichthyan response to, 116–119
  plasma Ca response to, 116, 117, 119, 120, 140
  plasma magnesium response to, 140, 188
  plasma phosphate response to, 117
  plasma sodium response to, 140
  seasonal response to in osteichthyans, 117–119
  scale response to in osteichthyans, 118, 119
Hypophysis, *see* Pituitary, *see also* Hypophysectomy

I
Inhibitors of calcification, 15, 16
Integument, *see* Skin, *see also* Gill
Inner ear, *see also* Endolymphatic sacs, 1, 35–38
  Ca deposits in, 1, 35–38
Intestine, *see* Gut
Iodine-125 and 131 labelling of parathyroid hormone, 50
Ion concentrations in environment, 6–10, 91
Ion exchange in calcified cartilage, 34

SUBJECT INDEX

Ionic activities of plasma, 18, 19
Isomorphic replacement, 13
Isopedin, 30

K
Keratin, 40
Kidney, see also Renal, 81–84
  and divalent ion regulation in cyclostomes, 95
  and vitamin D_3 metabolism, 64, 82
  role in chondrichthyan Ca regulation, 98

L
Lactation, 65, 66
Lactic acid
  response to calcitonin, 76
  response to parathyroid hormone, 75
  in bone resorption, 75
Lacunae
  in bone, 23, 25
  in cartilage, 35
Larval stage, 88, 123, 129, 141, 196
Life cycle stages in European eel, 100
Ligand formation, 14
Lime sacs, see Endolymphatic sacs
Lithosphere, composition of, 7, 8, 9
Liver, production of Ca binding protein by, 11, 70, 181
Lunar influences, 138, 139
Lysosomes, 20

M
Magnesium
  cycle in nature, 8, 10
  freshwater concentration of, 9, 10
  gut transport of, 80
  homeostasis, 7, 140, 187, 188
  interaction with Ca metabolism 187, 188
  role in calcification, 16
  in parathyroid hormone secretion, 48
  sea water concentration of, 6, 7
  skeletal content of, 22
  inhibition of calcification, 16
Magnesium: Ca, ratio in plasma, 16
Medullary bone, 31–34, 165, 167, 168, 169, 170, 181
  absence in reptiles, 34, 155
  as a Ca reservoir, 165
  cellular activity of, 33, 167, 168

chronic response to calcitonin, 170, 171
  formation in rodents, 78
  maintenance in egglay, 167–169
  organic composition of, 33
  rate of formation, 33
  resorption in egglay, 32, 33, 165, 168, 169
  response to calcitonin, 170, 171
  response to Ca deficiency, 168
  response to parathyroid hormone, 167–169
  role of gonadal hormones in formation, 33, 78
  sensitivity to parathyroid hormone, 168, 169
  surface area of, 32, 166, 168
Melanophore stimulating hormone, 120
Membrane bone, see Dermal bone
Mesenchyme cell as oesteogenic precursor, 21, 24, 26
Metamorphosis of amphibia, 123, 141
Mineralization, see also Accretion, 3, 12, 14–20, 26, 34
  chemistry of, 14–19
Monocytes as osteogenic precursors, 25
MSH, see Melanophore stimulating hormone

N
Neuronal supply to ultimobranchial gland, 130, 131
Nucleation sites, 15, 16, 20, 40

O
Ocean, evolution of, 6
Odontoblasts, 39
Odontoclasts, 39
Odontocytes, 39
Oestradiol and vitamin D_3 metabolism, 66, 174
Oestrogen
  and medullary bone formation, 33, 78
  chondrichthyan response to, 98
  effect on bone, 78, 121, 165
  effect on carbonic anhydrase, 179
  effect on fish scales, 31, 85, 121
  effect on mammalian bone, 78
  effect on vitamin D_3 metabolism, 66, 174
  effect on vitellin synthesis, 70, 71, 121

218 SUBJECT INDEX

Oestrogen—*cont'd*
  hypercalcaemic response to, 70, 71, 98, 121, 181
  osteichthyan response to, 121
  in medullary bone formation, 33, 78, 165
Open system of Ca regulation, 92, 93, 95, 97
Organic acids, role in bone resorption, 75
Ornithine cycle, 96
Osmolality of chondrichthyan plasma, 96
Osmoregulation, 83, 86, 92, 96, 106, 110, 113, 120, 147
Osteoblast, 24, 30, 32
  function of, 24
  in medullary bone, 32, 33, 168
  response to calcitonin, 76, 104
  response to parathyroid hormone, 74, 75, 168
  role in bone matrix secretion, 24
Osteoclast, 24, 25, 30, 32, 33
  function of, 25
  in fish scales, 30
  in medullary bone, 32, 33, 167
  in primitive bone, 25
  response to calcitonin, 76, 104
  response to parathyroid hormone, 74, 126, 149, 167
  response to Stanniectomy, 111, 113
  response to parathyroidectomy, 126
  response to ultimobranchialectomy, 132
  in bone resorption, 24, 25
Osteocyte, 23, 24, 28, 30
  function of, 24
  in fish scales, 30
  response to calcitonin, 76, 104
  response to parathyroid hormone, 74, 126
  in bone resorption, 24
Osteogenesis, *see* Bone formation
Osteogenic cells, 24, 25, 26
Osteolysis, 24, 74, 76
Otoliths, 12, 37, 38
Ovariectomy and vitamin D_3 metabolism, 174
Oviduct, *see also* Shell gland, 88, 89, 175, 176, 178, 179, 181
  Ca transport by, 88, 89, 175, 176

  as parathyroid hormone target organ, 165, 166
  Ca^{2+}/Mg^{2+} ATPase activity of, 88
  Ca binding protein of, 88, 89, 176
Ovulation effect on Ca appetite in hens, 179, 180

P

^{32}P-Phosphate as bone label in acid-base studies, 190
Paralactin *see* Prolactin
Parathyroid gland
  accessory tissue, 43, 157
  effect on gut Ca uptake, 134, 135, 137
  embryology of, 42
  histology in amphibians, 126, 127
  avian, 156–159, 166–169
  changes in size, 166
  morphology, 41–43, 149, 156, 157
  reptilian, 148–150
Parathyroid hormone, 5, 41–51
  active fragments of, 45
  amphibian response to, 125–129
  and bone carbonate mobilization in acidosis, 191
  and buffer base mobilization, 189–194
  and carbonic anhydrase in oviduct, 176
  and renal acid excretion, 190, 191
  and renal bicarbonate excretion, 190, 191
  avian response to, 160, 166–169
  bioassay of, 47, 48
  bone anabolic response to, 75
  bone response to, 74–76, 125–129, 167–169
  catabolic effects of, 74, 75
  chemistry of, 43–46
  circulating levels of, 49–51, 167–169
  degradation of, 50, 51
  diuretic response to, 159
  effect on acid excretion by toad bladder *in vitro*, 146, 191
  effect on Ca appetite, 179, 180
  effect on cellular Ca uptake, 74
  endogenous activity in hens, 167–169
  endolymphatic sac response to, 125, 145
  evolution of, 45, 70, 75
  fish response to, 75
  gut, specific response to, 79, 80

frog skin response to, 85, 128, 129
from avian glands, 45, 46
glandular levels, 49
initial hypocalcaemic response to, 74
isohormonal forms of, 44, 45
plasma phosphate response to, 149–150
phosphaturic response to, 75, 159
precursor of, see Pro-parathyroid hormone, Pre-proparathyroid hormone
radio-immunoassay of, 47
radio-immunoassay of, in acidotic subjects, 191
rapidity of response to, 47, 50, 51, 157–159
renal adenyl-cyclase response to, 159
renal response to, 82, 83, 125, 149, 150, 159
role in acid-base balance, 189–194
role in hibernation, 194
role in magnesium homeostasis, 188
secretion of, 48, 49
  Ca ion influence, 48
  calcitonin influence, 49
  3′, 5′-cyclic AMP role in, 48
  magnesium ion influence, 48
  influence of pH, 48, 49, 191
  vitamin D_2 influence of, 49
Parathyroidectomy
  acid-base response in amphibians, 143–145, 189
  amphibian response to, 123, 125, 127, 128, 131, 134, 135, 137, 140, 143–145
  avian responses to, 157
  effect on acid-base regulation, 191–193
  endolymphatic sac response to, 143, 145
  plasma magnesium response to, 140
  plasma phosphate response to, 148, 149
  reptilian response to, 148, 149
Pentagastrin, effect on calcitonin secretion, 58
Perilymph, composition of, 36
Peritoneal lavage
  parathyroid gland response to, 126
  studies in amphibians, 126
pH, see Hydrogen ion concentration

Pharyngeal pouch derivatives, 41, 42, 52
Phosphate
  and calcitonin response, 185, 186
  as a buffer-base reserve, 189
  homeostasis, 3, 4, 183–187
  in hard tissues, 3, 4, 7, 12, 21, 35
  plasma level in reproducing female, 154
  plasma response to parathyroidectomy, 125
  storage in chondrichthyan cartilage, 97
  transport by gut, 79, 81
  renal excretion of, 82, 83, 84, 168, 169
  response to Stanniectomy, 111
  urinary excretion of in acidosis, 141, 142, 191, 193
  urine response to parathyroid hormone, 125
  urine response to parathyroidectomy, 125
Phosphate, interaction with Ca metabolism, 183–187
Phospholiproprotein, see Vitellin
Phosphorus
  availability in environment, 3, 4, 6, 7, 9, 10, 183, 185
  bone as a store of, 3, 4, 53
  cycle in nature, 3, 4, 9
  dietary availability of, 3, 185, 186
  freshwater concentration, 9, 10
  homeostasis of, 3, 4, 7, 183–187
  in acellular bone, 29
  sea water concentration of, 6, 7
Phosphorus-32, studies in the bone response to calcitonin, 76, 184, 185
Phosphorylase kinase, 74
Photoperiod, influence on avian egglay, 174, 175
Pituitary gland, see also Hypophysectomy, 69, 70
  extirpation see hypophysectomy
  amphibian response to, 139, 140
  effect on fish scale turnover, 79, 117–119
  influence on bone, 78, 79
  interaction with calcitonin in fish, 104
  osteichthyan response to, 116–121
  role in magnesium metabolism, 188
  role in osteichthyan Ca regulation, 116–121

SUBJECT INDEX

Pituitary hormones
  and parathyroid hormone evolution, 44, 45
  and phosphate regulation, 187
  role in agnathans, 70, 95
Placenta, 88
Plasma calcitonin level
  in fish, 60, 61, 106–110
  in birds, 61, 62, 170
  influence of gonadal hormones on, 170, 186
  pre-laying surge of, 170
Plasma Ca
  complexed fraction, 10
  diffusible fraction, 10
  ionized fraction, 10, 11
  protein bound fraction, 10, 11
Plasma Ca level
  in acidosis, 192
  in amphibians, 123, 124
  in birds, 156
  in chondrichthyans, 96
  in cyclostomes, 6, 9, 91
  in humans, 6
  in osteichthyans, 9, 99, 100
  in ovulating hens, 169, 170, 171
  in reptiles, 147
Plasma magnesium level
  in reproducing females, 153, 154
  response to hypophysectomy, 140, 188
  response to Stanniectomy, 111
Plasma phosphate level
  in amphibians, 124
  in cyclostomes, 6, 9, 91
  in osteichthyans, 9
  in reptiles, 147, 154
Post-menopausal osteoporosis, 183
Potassium
  concentration in bone fluid, 22, 23
  concentration in endolymph, 36, 37
  content of bone, 22
  content of bone fluid, 22, 23
Potassium/sodium ratio in endolymph, 37
Precipitation of Ca salts, 13, 15, 16, 18
Preen gland, 63
Pregnancy, 65, 66, 88
Pre-pro-parathyroid hormone, 46
Pro-calcitonin, 57
Progesterone
  effect on calcitonin secretion, 58
  effect on carbonic anhydrase, 179
Prolactin, 69, 70, 97
  and Ca flux in fish, 120
  and urinary Ca excretion, 84
  and vitamin D_3 metabolism, 65, 66, 174
  bone response to, 79
  effect on water and sodium regulation in fish, 120
  pituitary levels of, in euryhaline fish, 120
  plasma Ca response to, 119, 120
  role in water and electrolyte regulation, 120, 195
Pro-parathyroid hormone, 46
Prostaglandin E, 83
Proteoglycan, 39
  in cartilage, 34
  in eggshell, 40
PTH, see Parathyroid hormone
Pyrophosphate
  inhibition of calcification, 16
Pyrophosphatase, 16, 76, 77

R

Radio-immunoassay, 46–48, 51, 57, 59, 106
Radioisotopes, see individual isotopes
Radio-receptor assay, 66, 67
Rectal salt gland, 97
Renal stones, 101, 117
Renin
  from the corpuscles of Sannius, 67
Resorption, see also Bone, resorption of
  definition, 14
  of hard tissues, 14, 16, 24, 25, 27, 30, 32
Respiratory acidosis
  and parathyroid hormone evolution, 194
Ribonucleic acid, see RNA
Rickets, 62
RNA, 46
  and Ca binding protein, 78

S

Sacculus, 35, 36
Salt concentrations, see Ion concentrations

SUBJECT INDEX

Scales
  as a labile mineral store, 85, 86, 187
  chondrichthyan, 29, 30, 96
  in tetrapods, 31
  osteichthyan, 103, 104, 118, 121
Sea water, composition of, 6, 7, 19
Seasonal cycle
  of C cells, 53
  of amphibian parathyroid gland activity, 126, 127
  of amphibian ultimobranchial activity, 130, 131
Seasonal fluctuation, of blood phosphate levels in amphibians, 124
Seasonal response to hypophysectomy, in fish, 117–119
Seasonal rhythms of Ca regulation, 26, 124, 126, 127, 131
Seasonal sensitivity of response to vitamin D_3, 134–136
Seasonal variations of plasma Ca levels, 147
Secretagogues, effect on calcitonin secretion, 58
Serum, see Plasma
Shell gland, 165, 166, 172, 175–179
Silica-based skeletons, 12
Skeleton
  evolution of, 1–4
  functions of, 1, 3, 4
  homeostasis of, 5
Skin
  amphibian, 85, 128, 129
  Ca exchange by, 84, 85, 95, 103, 104, 116, 118, 128, 129
Sodium
  content of bone, 22
  regulation of, 102, 103, 125, 131, 140
  renal excretion of, 82, 83, 125, 131
  urinary response to parathyroidectomy, 125
  urinary response to ultimobranchialectomy, 131
Sodium-22, flux studies in fish, 102, 103
Sodium adenosine 5'-triphosphatase, see Na^+ ATPase
Soft tissue as Ca reservoir, 40
Solubility product, 15, 18, 19, 75, 187
Somatotrophin, 5, 72, 119
Sodium [^{14}]-bicarbonate studies in acid-base regulation, 176, 177

Stanniectomy
  response to, 110–113
  urinary Ca response to, 111–113
Stannius corpuscles see Corpuscles of Stannius
Statoliths, 37, 38
Stenohaline fish, 92
STH see Somatotrophin
Strontium, 12, 163
Sulphur-35 studies in amphibian collagenous tissues, 128–129

T

TCT, see Calcitonin
Teeth, 29, 38–40
Temperature effect on amphibian response to parathyroid hormone, 126–128
Testosterone and vitamin D_3 metabolism, 174
Tetany, 69, 117, 125, 148, 149, 157
Thermodynamic ion product, 18, 19
Thyrocalcitonin, see Calcitonin
Thyroid gland, see also C cells, 5, 42, 43, 53, 54, 58, 186

U

Ultimobranchial C cells, embryonic origin, 52, 159
Ultimobranchial gland, 41, 42, 43, 52, 53, 54
  darkness, response to in teleost fish, 100, 101
  changes in size, 169
  culture *in vitro*, 162
  embryology of, 52
  extracts, assay of, 151
  innervation of, 130, 131, 162
  morphology, 43, 52, 53, 129, 130, 151, 159
  response to Stanniectomy, 111
  role in amphibians, 129–132
Ultimobranchialectomy
  acid-base response in amphibians, 143–145
  amphibian response to, 130–132, 135, 137, 143–145
  avian response to, 160
  bone response to, 132
  corpuscles of Stannius, histological response to, 105

Ultimobranchial—*cont'd*
  diuretic response to, 131
  eggshell thickness response to, 160
  endolymphatic sac response to, 143, 145
  gut Ca uptake response to, 134, 135, 137
  osteichthyan response to, 105, 106
  plasma chloride response to, 106, 107
  urinary Ca response to, 131
  urinary sodium response to, 131
Urea, 97
Urinary bladder, role in acid-base balance, 146, 191
Uronic acid excretion, 159
Uterus, *see* Oviduct
Utriculus, 35, 36

V

Vertebrates, classification of, 2
Vitamin A, 72
Vitamin C, 101
Vitamin D_2, 134
Vitamin D_3, 5, 62–67
  amphibian response to, 132–137
  and bone Ca uptake, 132–134
  and phosphate regulation, 186, 187
  avian response to, 163, 164, 172–176
  chemistry of, 62–65
  chondrichthyan response to, 98
  evolution of metabolism, 195
  in agnathans, 64, 95
  and endolymphatic sac Ca uptake, 132–134
  deficiency effect on parathyroid gland, 46
  deficiency in egg-laying hen, 172
  egg-laying hen response to, 172, 176
  endolymphatic sac response to, 132–134
  levels in fish liver, 114
  metabolism in birds, 163, 172–174
  metabolism in egg-laying hen, 172–174
  metabolism in osteichthyans, 114, 115
  metabolism in reptiles, 152
  osteichthyan response to, 114–112
  oviduct Ca binding protein, response to, 88, 89, 176
  precursors in skin, 62, 63
  production of Ca binding protein in gut, 80, 81
  role in Ca transport by oviduct, 88, 89, 175, 176
  role in gut phosphate transport, 81
  role in tetrapods, 195
  seasonal variation in amphibian response to, 134–136
Vitamin D_3 metabolites, *see also* 25-hyroxycholecalciferol, 1, 25-dihydroxycholecalciferol, 24, 25-dihydroxycholecalciferol, 62–67
  carrier proteins for, 66
Vitellin, 70
Vitellogenesis, 70, 121, 152–154, 181

X

X-ray studies
  of bone, 132, 133
  of endolymphatic sacs, 125, 132, 133, 154

Y

Yolk proteins and Ca binding in hen, 165
Yttrium-91, 175